# Probiotic

# Cities

*Probiotic Cities* covers a body of work that is at the forefront
of emerging knowledge within architecture, towards designing
informed indoor and built environment microbiomes. Sited within the
broader field of Bio Design, the book presents highly experimental
design research at the intersection of architecture, engineering
and microbiology. The book describes work which explores novel
strategies towards directly (re)introducing beneficial microbes
into buildings and cities. Through discussion of both the work and
the processes and methodologies used, it provides a framework
to enable designers and practitioners to begin to engage with
contemporary human–microbe relationships towards the design
of healthy and resilient cities. The book defines a new microbial
paradigm for architecture that engages with broader emerging
ecological or 'more than human' philosophies for design within the
age of the Anthropocene.

**Richard Beckett** is an architect and Associate Professor of
Bioaugmented Design at The Bartlett School of Architecture,
University College London (UCL). He is Director of Research Cluster
7 on the BPro Architectural Design master's programme, and leads

Studio 3 on the Landscape Architecture course. His research is focused on design operating at the intersection of computation, biofabrication and microbial ecologies in buildings and cities. His research on probiotic design won the RIBA President's Research Award in 2021. He has built numerous projects and has been exhibited internationally, including at Archilab – 'Naturalising Architecture', the Pompidou Centre and at Cooper Hewitt, Smithsonian Design Museum – 'Nature'.

# Bio Design

*Series editor: Martyn Dade-Robertson,*
*Newcastle University, UK*

The Bio Design series offers the opportunity for designers from
fields as diverse as architecture, fashion design and product design
to present and explore designs and design research which use living
systems as part of their production and operation. The series offers
readers in-depth project descriptions, analysis of processes and
the intellectual contexts of Bio Design. Such explorations have not,
until now, been made available in long form. The series also allows
designers to explore the potentials and challenges of bio design as
an emerging field of design and research.

The Bio Design book series will distinguish between key areas
within the field, including, design fictions, biomimicry, bioinspired
design and the use of biological materials and systems. While open
to a range of voices the series will, as a collection, offer unified
framework for thinking about Bio Design opening up this new rapidly
growing field to a new generation of designers and researchers.

# Probiotic Cities

**Richard Beckett**

Routledge
Taylor & Francis Group

LONDON AND NEW YORK

Designed cover image: © Richard Beckett

First published 2024
by Routledge
4 Park Square, Milton Park, Abingdon, Oxon OX14 4RN

and by Routledge
605 Third Avenue, New York, NY 10158

*Routledge is an imprint of the Taylor & Francis Group, an informa business*

*British Library Cataloguing-in-Publication Data*
A catalogue record for this book is available from the British Library

ISBN: 9781032076096 (hbk)
ISBN: 9781032076102 (pbk)
ISBN: 9781003207917 (ebk)

DOI: 10.4324/9781003207917

Typeset in Calvert
by codeMantra

# Contents

# Series editor's preface

The emergence of biotechnologies and their integration into human-centred contexts of use has led to a new design paradigm known as Bio Design. Bio Design, bio-design or biodesign, depending on the source, has a wide range of definitions. Early examples of Bio Design are mainly focused on the design of medical devices and methods to support tissue engineering. Bio Design has also been used alongside fields such as synthetic biology. This series, however, uses a broader definition of Bio Design which includes the integration of biological processes across a range of creative design fields including, but not limited to, architecture, fashion and apparel design, product design and interaction design. Too often, real applications and technologies are confused with grandiose pronouncements of technological revolution based on science fictions built on limited or no real engagement with messy biological reality. The series will allow for deeper explorations of Bio Design projects and theories. The books will necessitate critical and reflective descriptions of the Bio Design field focused on the underlying processes, methods and theories. In defining the territory of this book and the series it initiates, we are seeking a happy middle ground between design speculation and grounded experiment, between critical thinking and creative naivety, between formal elegance and radical complexity and to provide ways of thinking which will lead to critical ways of making.

Martyn Dade-Robertson
Newcastle University
February 2020

# Ethics and society

Any conversation on designing with biology necessitates refection about ethics. Novel biotechnology and bioengineering applications have the potential to provide enormous benefits for society. However, the relationship between human activities and the current climate emergency due to global warming reminds us that technologies used to solve one set of problems are capable of creating many others. Beyond environmental impacts, there is also the challenge of considering diverse political and social values in the development of bioengineered technologies and materials. As a result, we wanted to establish an ethical position in developing this series.

- We will not seek to publish material which could be applied directly to the development of weapons or deliberate causes of harm.

- Where experiments or processes are introduced which may be harmful to the individuals conducting them or the environment, we will make these risks clear, especially given audiences who may be unfamiliar with the techniques and technologies being described.

- Authors in the series will be required to confirm (where applicable) that appropriate risk assessment, ethics review, informed consent and animal welfare protocols have been met, in compliance with local institutional and governmental regulations.

While we make every effort to anticipate risks in our research, the unintended consequences of technology are harder to predict.

However, in formulating this series, we believe that Bio Design is at its best when it is a reflective and inclusive practice where ethical and responsible principles are embedded. The books and editorial guidance in this series prioritize this, while accepting diversity of opinion and position.

Martyn Dade-Robertson
Carmen McLeod
Newcastle University
February 2020

# Acknowledgements

When I made the decision to retrain as an architect, I left my career as a scientist with the desire to follow a more artistic, creative path, swapping the laboratory for the design studio. This worked well during my undergraduate degree. It was when I undertook my Master's degree at The Bartlett School of Architecture at University College London (UCL) that I was encouraged to utilise my previous background and I found myself back in the laboratory, now exploring a more interdisciplinary approach towards architecture and engaging in the field of biodesign. The experimental agenda of The Bartlett has been fundamental to this research both in my time as a student and as an academic, as has UCL in facilitating cross-disciplinary research. I would like to thank Marcos Cruz, with whom I co-directed the BiotA Lab, for numerous discussions on architecture and biodesign along with other colleagues in the field, including Martyn Dade-Robertson and Carolina Ramirez-Figueroa, who have been fundamental in shaping this approach. Thanks also to Sean Nair (UCL), David Thaler (University of Basel) and Jack Gilbert (University of California, San Diego) for fielding many of my microbe and microbiome questions and embracing these ideas with intellectual collaboration.

All of the projects described in this book have been supported through research funding. They have also been undertaken with a series of collaborators, underpinning the interdisciplinary nature of the experimental work. To acknowledge these collaborations, I use 'we' when discussing the projects. The Niches for Organic Territories in Bio-Augmented Design (NOTBAD) project was funded by the Arts and Humanities Research Council (AH/R001987/1). The microbiological work was undertaken in the laboratories at the Eastman Dental Institute, UCL, under the supervision of Sean Nair. The experiments were conducted with Carolina Ramirez-Figueroa, who worked on this research with me, and the final experiments towards MRSA inhibition were undertaken by Mehmet Devrandi and supported by others in Dr Nair's lab. The meso-scale work in

Chapter 4 was funded partly by The Bartlett Architectural Research Fund (ARF) and partly by UCL Grand Challenges. My collaborators here were Lena Ciric from Civil, Environmental and Geomatic Engineering at UCL, along with Darren Smith and Greg Young from Northumbria University, who led the genomic sequencing work. Finally, in the macro-scale work, funding from UCL Grand Challenges facilitated further collaboration with Sean Nair and Yang Gao, and an industrial collaboration with Incremental 3D towards the development and fabrication of the prototype. Thank you to all involved.

I would also like thank my Master of Architecture students from Research Cluster 7, of whom there are too many to name but who also helped shape the ideas in this book. Some of their work is shown in the book, and paves the way for ongoing work in the coming years as we seek to form new networks and collaborations with clinicians and immunologists to shape healthy and resilient cities.

Finally, I want to thank and dedicate this book to my family: to Emma, Amelie and Sterling, the loves of my life.

# Introduction

## Chapter 1

DOI: 10.4324/9781003207917-1

## BIO-INTEGRATED DESIGN

In 2014 I co-founded the Biotechnology and Architecture Lab (BiotA Lab), which was set up as a research group at UCL's Bartlett School of Architecture to explore new approaches within the emerging biodesign field seeking to find ways to integrate living biological species, materials and systems into architecture.[1] Underpinning the design approach at this time was an attempt to reconceptualise the popular metaphor of the building envelope from that of 'skin' to that of 'tree bark'. The work developed a new 'bioreceptive' design paradigm (Cruz & Beckett, 2016), building on this notion of tree bark, where the envelope becomes one which is both protective and productive, serving as a host for biological growth of other species. The group focussed on developing bioreceptive materials which, as a result of their design and fabrication, could be optimised in terms of their physical and chemical properties to support the growth of photosynthetic microorganisms such as algae, lichen and mosses directly on the substratum of the material. We were given a kitchen space within the university, which was the closest we could get to a wet lab, with some basic laboratory equipment. The working space became part design studio, part laboratory, part workshop. The students worked in a hybrid manner, combining laboratory methodologies with design research, computation and fabrication to explore new ways of integrating nature into the very material of architecture.

We positioned this work at this time as a novel approach to existing green or plant wall systems that, though popular, were seen to be costly to implement and intensive to maintain. Others in the field were exploring more biotechnological approaches to greening building façades, growing liquid cultures of algae and cyanobacteria in photobioreactors attached to buildings (Wurm, 2013). An important part of our approach at the time, I believe, was a rejection of the 'container' as the biological niche for large-scale application on buildings. The direct application of the microbiology lab on to a building was seen as inherently problematic, primarily due to high costs and the need for intensive ongoing maintenance. Whereas plastic or glass containers in the sense of large petri dishes or bioreactors serve well for monoculture growth in controlled environments like laboratories, maintaining monocultures at large scales, in real-world environments such as cities, is challenging, energy-intensive and costly. We also questioned the concept of 'containing' nature in closed vessels, which we saw to be fundamentally separating the biological matter and agency not only from the architecture itself but also from the occupants and the surrounding ecologies.

Conceptually, this work was exploring instead the notion of 'integration' of biological matter into the architectural fabric, and, ultimately, we were designing for conditions of 'exposure'. The relationships and dynamics between the living organisms, the material surface and exposure to the environment became important. Environmental conditions of sunlight, shade, moisture and wind became integral metadata parameters informing the computational design of the three-dimensional surfaces and the microclimatic conditions they created to support (or inhibit) biological growth. Although we were targeting the visible growth of mosses, it was understood that this would never become a pure or monoculture condition. The succession of viable moss colonies would be dependent on a whole ecosystem underpinning it, including other species – pioneer microbes (such as algae and bacteria) as well as a host of other organic and non-living particles and molecules integrated in the material.

**FROM OUTDOORS TO INDOORS**

While this work was mostly aimed at outward-looking, urban greening strategies, more recently my interest in this research shifted the focus of these bio-integrated design approaches to also include indoor environments. In doing so, I began to consider the roles and agencies of microorganisms inside buildings. It became less about green, photosynthetic organisms. The indoor condition became more about other, less visible microbes – bacteria, archaea and even viruses. It saw a shift in the agenda of this work from a predominantly climatic discourse to one concerned with health and human–microbe relationships in architecture. The research agenda began to engage with the microbiome of the built environment (MoBE). Currently, no roles exist for microbes inside buildings. In fact, since the medical understanding that microbes caused disease became widely accepted around the turn of the 20th century, architecture has been increasingly employing strategies to remove and eradicate microbial life from inside buildings. The modern city, in this way, has been defined on the assumption that healthy buildings = fewer microbes.

While much is known about the negative relationships between microbes and health, my research began to engage with the emerging, contemporary understanding of microbes and the beneficial roles they play in bodies and environments. The last 15 years have seen fundamental shifts in our knowledge of microbes and their evolutionary mechanisms. Technological advances in microbiome science have revolutionised our understanding of microbes and their importance within biological systems, including human health. Unlike in the past, we now know that not all microbes are pathogens; most are benign and many are beneficial, even essential for health. Disease is now understood to be more than just microbes (Hinchliffe et al., 2016); in fact, the body is dependent on exposure to microbes for normal healthy function. A contemporary reading of health and human–microbe relationships across the sciences and humanities fundamentally challenges the modern assertion that, to make life safer, the human should be separated from non-human life. More pressing is emerging evidence that shows how antibiotic attempts to create increasingly sterile built environments that favour strategies of eradication and separation are resulting in new, unintended and even more challenging microbial pathologies (Blaser, 2014, p. 288). These unintended conditions that are detrimental to health form an integral part of this probiotic design approach.

## A PROBIOTIC SHIFT

I began to question what it would mean to challenge broader existing approaches to architecture that were 'antibiotic' in their philosophy, and instead sought to consider what an opposite, probiotic approach might mean. If strategies that seek to collectively eradicate microbes are in fact creating unhealthy built environments through a lack of microbes, perhaps a new approach is needed that facilitates microbial presence in buildings and promotes exposure to them. If we currently configure buildings in line with the image of all microbes as pathogens, how might we configure buildings based on the understanding of the symbiotic roles microbes play in bodily health? This raises the challenging idea that, to create healthy buildings, we might need to design for more microbes, not fewer.

This thinking has not developed in isolation, and this work sites itself as part of a wider movement across a range of disciplines that understand and are now prioritising the fundamental role that macro- and microbiodiversity play as part of healthy and resilient environments. This wider movement, or 'probiotic turn', is identified by Jamie Lorimer in his book *The Probiotic Planet* (2020), where he distinguishes the term 'probiotic' from its popular association with live yoghurts or supplements. Instead he uses it in a broader, more expansive manner to describe ways of managing life that use biological processes to deliver health. These examples are presented in opposition to the antibiotic mindsets described above. They are seeking to restore or rebalance states of dysbiosis caused by modern antibiotic strategies. Again used in a broader sense than a class of medicines, Lorimer describes being 'antibiotic' through attempts to secure life that have favoured mechanisms of eradication, control, rationalisation and simplification. While these approaches have certainly done much good, their overuse has led to conditions of 'antibiotic blowback' including degradation of landscapes, extinction of species, conditions that facilitate disease emergence and states of microbial dysbiosis.

Probiotic approaches instead use 'life to manage life'. Lorimer (2017) offers examples of rewilding projects in the fields of landscape and ecology that seek to reintroduce ecologically important species to environments in order to restore biodiversity or enhance healthy ecosystem functionality. At the scale of the body he offers examples such as the medical shift towards gut biome restoration through emerging relationships between microbes and immune function. Probiotic mentalities are based on the sciences of immunity and ecologies, but advocates are also radical in their approaches. Some speculate with genetic engineering to bring back extinct species such as the woolly mammoth for their ecological agency, while others self-treat with hookworms to train and restore normal immune function (Lorimer, 2017).

When considered through the lens of architecture and the built environment we can observe a similar situation. The modern city embraced notions of hygiene and cleanliness that solved many of the health challenges associated with the industrial city, which was pathological through its filth and abundance of pathogens (Drake, 1997). However there is a sense that, at some point, these approaches have gone too far and we may have now passed a tipping point. In being overly antibiotic, the contemporary city is now being characterised as unhealthy through its sterility and lack of microbial diversity (Sariola & Gilbert, 2020). Increasingly

Introduction

sterile indoor environments, synthetic materials and overuse of antimicrobial cleaning products are driving the selection and evolution of microbes towards more challenging and resistant states. Urbanization, loss of biodiversity and increasing confinement of indoor environments from the outside are being attributed to the rapid emergence of chronic and autoimmune diseases in urban populations (Velasquez-Manoff, 2012). From a broader perspective we see how these antibiotic mentalities are equally implicit in many of the contemporary climate challenges and the destructive impact of the Anthropocene.

## PROBIOTIC DESIGN

Facing such challenges in the built environment has typically been met with a 'more of the same' response, where a belief that more technology and new antibiotics can solve these problems. The response to the Covid pandemic highlighted how antibiotic mindsets remain predominant in the built environment. Buildings became framed in defensive terms in line with the warfare narrative that exists between humans and microbes. Homes became described as 'bunkers', separated from the unsafe and unruly outside. Public spaces were intensively sanitised, and buildings became accessible only to those who could prove they were free of microbes. The evolutionary mechanisms of microbes suggest this is a war that can never be 'won' and will only exacerbate chronic diseases associated with a lack of microbes. Others speculate on looking backwards, with ideals of pre-modern, even prehistoric lifestyles as the solution. Within the popular media there is a conception that we need to somehow live less clean or 'dirtier' lives. The probiotic approach developed here does not reject modern hygiene. Indeed, sanitation and hygiene are crucial to reducing the risk and spread of infection. However understanding how to design and plan our built environments in a manner that protects us from harmful microbes but permits and facilitates beneficial microbial exposures is crucial.

Probiotic design (Beckett, 2021) is an experimental research approach that begins to address these questions through radical new design processes to inform new visions for future cities that are microbial, healthy and resilient. It draws on the contemporary understanding of the evolutionary and symbiotic relationships that exist between microbes, human hosts and the built environments we inhabit. It engages the sciences of the microbiome, both of the human and of the built environment, with more social, spatial and aesthetic considerations embedded in architectural knowledge. To begin to plan and design buildings that host diverse, beneficial microbial communities, new design methodologies are needed that can engage with the sciences of the indoor microbiome and the medical fields of immunity. Probiotic design incorporates new networks of expertise bringing together architects, experts in infectious diseases, immunologists and environmental microbiologists. It engages design approaches with 'omics' tools such as 16S rRNA gene sequencing and metagenomic approaches that frame buildings through the concept of the microbiome. These offer new concepts of buildings and bodies, understood instead through code and big data sets of genomic information. These can be quantitatively analysed not only to distinguish specific strains and species but also to frame spaces through the complex and evolutionary entanglements between human and non-human life.

The majority of the knowledge of the MoBE currently lies within the fields of medical and building sciences. To date, this has focussed mostly on its characterisation (National Academies of Sciences, 2017). What is clear is that architectural design plays a major role in shaping the constitution of the MoBE. What emerges too is the potential for architecture to begin to inform and shape the MoBE, particularly the indoor microbiome (IM), towards a more healthy condition (Kembel et al., 2014) by increasing the presence of microbes in buildings. Yet the challenge is more than just a technical one; the additional presence of microbes in buildings is unlikely to be enough. Probiotic design also considers the notion of microbial exposure that occurs via mechanisms of touch, respiration and ingestion. These have socio-economic as well as physical and spatial considerations. Here the work looks to how design can serve to reframe these as beneficial microbial entanglements. It explores how buildings might be shaped and inhabited in new ways to promote a thickening of the microbial contact between the building, the dermal layers, the nasal passages and the gut of holobiont bodies. To do this, it is necessary to incorporate strategies, materials and species that are currently undesirable, even taboo, in architecture. It is also necessary to challenge strategies and aesthetics of sterility and engage instead with porous materials, moisture, dirt and ageing, and their resultant, uncontrolled aesthetics.

## A NOVEL DESIGN CONTEXT

This probiotic design agenda offers new and important lines of enquiry across the broader fields of biodesign, bio-integrated design, living construction and biofabrication that can be considered through the lens of the holobiont body and the context of the microbiome and immunoregulatory health. Architects operating at the intersection of design and biology would be well placed to contribute to this agenda. Designing with biological species to create materials for buildings is common in the broader 'living architecture' discourse. Yet while the more-than-human zeitgeist has driven the focus of designers towards non-human species, when applied predominantly through the context of climate, existing bio-integrated design approaches have rationalised and favoured specific photosynthetic species at the expense of wider, non-green diversity. For these approaches to be applied to buildings, there has been pressure to try to quantify their productivity on a per $m^2$ basis for them to be considered viable. When the aim of these approaches turns to productivity, strategies of control, rationalisation and simplification tend to re-emerge.

## SCOPE OF THE BOOK

This book seeks to make the research area of the MoBE and holobiont health visible to architects and designers interested in the field of biodesign who are engaging with experimental approaches using life that is necessary to imagine and design future cities that are, like the human holobiont, fundamentally microbial. I hope that it can provide new lines of enquiry beyond the 'green' climatic and sustainability agenda that has dominated the field to date. I hope it also helps encourage experimental approaches not only towards new ways to design buildings but also to explore how we construct and inhabit them through the role they can play in informing beneficial microbial entanglements.

The way in which I have approached the work has attempted to keep the research relevant for architects and designers operating at the interface of architecture and biotechnology. This is probably not the way a scientist or building engineer would approach the work. It is grounded in the existing science, yet it also makes leaps – speculative but informed moves both in scale and in aesthetics that seek to engage the creativity of architecture within a currently science-dominated field. It has tried to avoid using materials that are overly expensive to procure or to make, such as high-purity ingredients from scientific suppliers that are either hugely expensive or only available in small quantities. This tends to work well in engineering approaches under strict conditions of laboratory experimentation, but become less relevant when working at the scale of building materials, especially when the scale of the work goes beyond what is feasible in a laboratory. These approaches also make it difficult for others in the biodesign fields who wish to build on this approach but who may not have access to large amounts of funding to engage with such materials.

**BENEFICIAL MICROBES, INFORMED MICROBIOMES**
To be clear, it is still not known what constitutes a healthy human microbiome or which specific microbes might be necessary to constitute a healthy indoor microbiome. This book will therefore address the question of how we might start to design informed indoor microbiomes. Probiotic design differentiates between pathogenic and benign microbes, but also considers that microbes may not be inherently good or bad. Instead it understands that such a distinction may be configurable by the relationship between bodies, microbes and their environmental context. In this way it does not fully reject antibiotic approaches; instead it questions how we might recalibrate them using probiotic approaches that continue to limit harmful microbial exposure but crucially allow for and facilitate other benign and beneficial exposure.

This book will present two approaches that seek to use benign microbes to target healthier indoor microbiomes. The first will describe the work undertaken in the Niches for Organic Territories in Bio-Augmented Design (NOTBAD) project, which explores the potential to make indoor environments safer by reducing the presence and persistence of pathogens on building surfaces using beneficial bacteria. The second approach seeks to create more beneficial indoor microbiomes in line with the symbiotic relationship between humans and microbes to inform experimental approaches that seek to re-deploy diverse environmental microbial communities in indoor environments. These approaches aim to secure the desired ecological functions of diverse environmental microbes through built interventions that create ecological conditions and biological niches that support their presence in buildings and augment their flourishing in line with the human microbiome of building occupants.

## STRUCTURE OF THE BOOK

Following this introduction, Chapter 2 will present a broad overview of both the scientific and philosophical basis for a probiotic design shift for buildings, and discuss what probiotic mentalities might mean for architecture. It will frame these against the contemporary city which is problematic due to a lack of microbes resulting from over 50 years of antibiotic mentalities in architecture. I will discuss the emergence of these mentalities alongside the physiological and philosophical distinctions between what is human and non-human that underpin them. I will then use contemporary microbiome science and the emergence of contemporary microbial pathologies associated with missing microbes to show that these distinctions are no longer relevant, in line with the contemporary understanding of the human as a holobiont body and the symbiotic roles that exist between humans and microbes. This contemporary position will inform some of the approaches presented in the rest of the book, rejecting indiscriminate antibiotic strategies and introducing instead design approaches underpinned by probiotic mentalities. The final three chapters will explore this agenda by presenting a series of novel probiotic design experiments and interventions that explore how to incorporate and rewild beneficial microbes into buildings to inform healthy indoor microbiomes.

These interventions are presented through design operations at three fundamental scales. Chapter 3 will describe methodologies operating at the micro-scale of the material–microbe interface to create novel living materials that are beneficial for the indoor microbiome through their ability to inhibit pathogens while, crucially, still allowing other benign microbes to remain. In Chapter 4 the design focus shifts upwards to the meso-scale of the material–building interface to explore how probiotic design can begin to engage and interface with the indoor microbiome. This is explored through interventions that seek to (re)introduce microbial diversity to buildings that serve as a source of beneficial microbes to the holobiont body. Finally, Chapter 5 discusses probiotic design approaches at the macro-scale of buildings and their microbial relationships with the broader urban environmental microbiome. This chapter will explore ways to configure urban environments and buildings in relation to the environmental microbiome so as to promote and enhance building and human entanglements with outdoor microbiodiversity towards shaping healthy and resilient probiotic cities.

**NOTE**

1  The BiotA Lab was co-founded by Marcos Cruz and Richard Beckett, along with Javier Ruiz, and ran from 2014 to 2018. The lab integrated students from Research Cluster 7 (RC7) on the UCL B-Pro, Architectural Design MArch programme.

**REFERENCES**

Beckett, R. (2021). Probiotic design. *Journal of Architecture*, *26*(1), 6–31.

Blaser, M. J. (2014). *Missing microbes : how killing bacteria creates modern plagues*. Simon & Schuster.

Cruz, M., & Beckett, R. (2016). Bioreceptive design: a novel approach to biodigital materiality. *Architectural Research Quarterly*, *20*(1), 51–64. https://doi.org/10.1017/S1359135516000130.

Drake, S. (1997). The architectural antimephitic: modernism and deodorisation. *Architectural Theory Review*, *2*(2), 17–28. https://doi.org/10.1080/13264829709478316.

Hinchliffe, S., Bingham, N., Allen, J., & Carter, S. (2016). *Pathological lives: disease, space and biopolitics*. John Wiley & Sons.

Kembel, S. W., Meadow, J. F., O'Connor, T. K., Mhuireach, G., Northcutt, D., Kline, J., Moriyama, M., Brown, G. Z., Bohannan, B. J. M., & Green, J. L. (2014). Architectural design drives the biogeography of indoor bacterial communities. *PLOS ONE*, *9*(1), e87093.

Lorimer, J. (2017). Probiotic environmentalities: rewilding with wolves and worms. *Theory, Culture and Society*, *34*(4), 27–48. https://doi.org/10.1177/0263276417695866.

Lorimer, J. (2020). *The probiotic planet: using life to manage life*. University of Minnesota Press.

National Academies of Sciences, Engineering, and Medicine (2017). *Microbiomes of the built environment: a research agenda for indoor microbiology, human health, and buildings*. National Academies Press.

Sariola, S., & Gilbert, S. F. (2020). Toward a symbiotic perspective on public health: recognizing the ambivalence of microbes in the Anthropocene. *Microorganisms*, *8*(5). https://doi.org/10.3390/microorganisms8050746.

Velasquez-Manoff, M. (2012). *An epidemic of absence: a new way of understanding allergies and autoimmune diseases*. Simon & Schuster.

Wurm, J. (2013). Developing bio-responsive façades: BIQ House – the first pilot project. *Arup Journal*, *2*, 90–95.

# Architecture for the Holobiont

## for the

## Holobiont

### Chapter 2

DOI: 10.4324/9781003207917-2

The Anthropocene is positioned to mark the end of a period of certainty that mankind's triumph over nature would secure the human. This is particularly evident in relationship to architecture, health and microbes. Following a golden age of architecture and health in the early 20th century, the development and success of antibiotic drugs in the mid-20th century brought with it a sense of confidence that humans had finally overcome the threat of infectious diseases once and for all. The medical idea that health should be based on the absence of microbes became embedded as accepted knowledge and these antibiotic mindsets permeated into architectural thought, shaping built environments that have sought to make human lives safer and healthier through the eradication of non-human threats to life. In some cases these approaches have done much good; but there is a sense that they have gone too far.

In the first decades of the 21st century, the world has been seen in perhaps its most infectious state ever. Not only have attempts to eradicate microbes failed, it seems they have also unintendedly created new microbial pathologies and even more challenging microbial environments. Over the last 50 years, antibiotic mindsets have resulted in the removal of nearly all unplanned and uncontrollable life from buildings and cities to the point where urban environments are now some of the least biodiverse habitats on the planet (Muller et al., 2010). Urban dwellers now spend up to 95 per cent of their day in indoor environments, where this lack of biodiversity, especially at the scale of microorganisms, is beginning to have detrimental effects on human health that we are only recently starting to understand.

These approaches are based upon a distinction between which forms of life are deemed to be good or safe and which are bad or not safe for humans. We will read how microbes, in line with the germ theory of disease, became diagrammed as linearly and collectively pathogenic, and so their elimination became the basis for creating healthy cities. But these mindsets were also informed by the modern conception of the human that was positioned as an island – a discrete body that was biologically distinct, and so one that should be physically and philosophically separated from other non-human life. In this chapter I want to address the distinctions we make between humans and non-humans, healthy insides and unhealthy outsides, and to assess the implications of these ontological distinctions for design and for a probiotic design approach to architecture. I will begin by tracing the emergence of antibiotic mentalities in architecture to make visible how they aligned with the modern medical diagram of microbes. I will then expand further on the recent scientific and ontological shifts offered by contemporary microbiome science that challenge these positions and assumptions. I will use the contemporary pathologies of antimicrobial resistance (AMR) and diseases of missing microbes to underpin the urgent need for a recalibration of how we shape our built environment in relation to microbes.

## ARCHITECTURE AND DISEASE

Architecture and health have a long and convoluted history. In tracing the ongoing dialogue between them, architectural mindsets have often shifted and aligned with the medical understanding of disease of the time to inform how to shape safer cities. Prior to the 19th century, beliefs about the causes of disease varied, and shifted from moral and astrological causes in the 17th century to ideas around contagion, infection and hereditary and environmental factors in the 18th century (Ozonoff, 1982). Without a solid theoretical base, disease was mostly understood as a behavioural issue, a punishment for sin or immoral behaviour (Magner, 2002), and so had little impact on architecture and the city. The major change can be observed in the second half of the 19th century, when responsibility for disease shifted from divine retribution to social causes, which eventually saw smells or 'miasmas' become the suspected causes of disease. For the first time, under the miasma theory of disease, we see health become related to the built environments the body inhabits; and we begin to see the emergence of architectural approaches that sought to improve health by removing these threats from the built environment.

## ARCHITECTURES OF SMELL

The eradication of smells and dirt from homes and streets revolutionised the health of people living in cities (Drake, 1997). Miasma theory, combined with efforts of social reformers and the emerging sciences of hygiene, saw architecture become medicalised. In response to the squalor and morbidity of the industrial city, unhealthy buildings were 'diagnosed' – not by architects but by physicians, who then 'healed' buildings using scientific systems of domestic sanitation. Strategies of cleansing to remove or mask miasmas eventually began to inform visual notions of cleanliness, and the architectural aesthetic at this time becomes similarly medicalised (and the medical aestheticised in architecture). Here, according to Mark Wigley (2001, p. 5), architectural cleanliness 'joins the doctor's white coat, the white tiles of the bathroom, [and] the white walls of the hospital'. Techniques of plastering, coating and whitewashing became common strategies that served to present visible cleanliness and aesthetics of 'impermeability' which kept bad smells off buildings but also exhibited the purity of water through the use of ceramics, varnishes and glazing (Carter, 2007).

By the turn of the 20th century, architecture and health were so intertwined that architects held the notion that buildings themselves could serve to cure disease (Aalto et al., 1998). Infectious diseases of the time, such as tuberculosis and rickets, informed the development of new building forms, spatial typologies and technologies that prioritised exposure to nature as the basis for health. This included design approaches to facilitate exposure to sunlight (heliotherapy), based on the belief that it could cure infections but also on its representation of purity. These aligned with the desire for clean, fresh air within buildings, a strategy informed from observations in hospitals, where wards with open windows demonstrated improvements in patient health and quicker recovery from illness compared to wards with no or closed windows. Some of the most prominent architectural experiments of this agenda included open-air schools and tuberculosis sanatoriums, where notions of healthy buildings were centred on exposure to nature.

This short history summarises the discourse of health in architecture prior to the widespread understanding that microbes caused disease. Although based somewhat incorrectly on miasma theory, these approaches measurably helped to reduce death rates from infection; and, along with the hygiene movement, life expectancy in cities increased significantly at the time. This trend can be seen to be well underway, even before germ theory was accepted by the medical field and the later development and widespread use of antibiotic medicines (Armstrong et al., 1999). Proper sanitation and hygiene measures would have reduced exposure to harmful microbes associated with the gut and faecal matter. The emerging social preference for cleanliness, both aesthetic and through material finishes, would have reduced the likelihood of harmful microbial succession on surfaces and materials in buildings, especially in spaces with moisture. Building strategies for maximising 'light and air' would have never been likely to cure infectious disease as hoped; however, they would have helped reduce the risk of infection by 'diluting' the presence of pathogenic microbes in the air and on surfaces in buildings (Hobday & Dancer, 2013).

**THE ANTIBIOTIC TURN**
Although these strategies would have served to remove microbes from buildings, I make the case that these approaches might not be considered antibiotic in their mentalities here. Primarily this is because concepts of hygiene and cleanliness at this time were not as conflated with the notion of sterility as they became later in the century. It marks then an important distinction between cleanliness, hygiene and sterility that is crucial in defining a probiotic design approach. The concept of hygiene at this time was tied to the maintenance of good health; yet these strategies were not driven by the broad spectrum removal of microbes or sterility as the target condition.

These same strategies were important in the recent Covid pandemic, where opening windows (in buildings that permitted it), hand and face hygiene and distancing measures were fundamental in reducing the spread of the virus as part of a public health driven 'targeted hygiene approach' in the UK (Hands, Face, Space). We also saw, however, examples of what has been described as 'hygiene theatre', which makes visible the blurred understanding between cleanliness and hygiene that predominates today (Rook & Bloomfield, 2021). These examples are presented as strategies that operate under the mistaken belief that they will protect against infection, but are in fact more likely to have adverse impacts on health. Practices such as the 'deep' or 'intensified' antibiotic cleaning of environmental surfaces, disinfection tunnels and fogging of buildings that sought to make buildings 'Covid-secure' have been described more in line with publicity and peace of mind rather than being effectual. So, when and why did this conflation between cleanliness and hygiene appear?

Despite death rates from infection falling at the turn of the 20th century, Tomes (2000) describes an explosion of 'antisepticonciousness' around this time. Medical understanding of infection was shifting away from miasmas towards germs. Tuberculosis, despite being on the wane, was used to popularise germ theory by reformers, underpinned by the new science of bacteriology. Public awareness of germs spread quickly due to new forms of mass media and advertisements that saw germ panic become a mechanism for selling new products promoting healthy ways of living with microbes. Germ theory had profound implications for the medical fields, and the focus on microbes and disease was translated away from the city as the problem towards that of a science problem to be solved through medicine and new technologies. This shift from urban reform to the laboratory sciences of bacteriology and medicine exacerbated the medicalisation of architecture at this time and had important implications in shaping antibiotic mindsets.

## HUMAN/NON-HUMAN DISTINCTIONS

What emerges from this shift is the so-called 'linear' understanding of microbes, where all microbes became collectively and solely diagrammed as causative agents of disease. The new field of bacteriology relied predominantly on culturing-based techniques; and, as pathogenic microbes became the subject of intense study, those that could not be attributed to disease (benign microbes) or could not be cultured (many microbes are difficult to grow in a laboratory) received little attention. Laboratory-based culturing methodologies saw a focus on single, isolated strains of bacteria and their behaviour in the petri dish. These techniques were unable to reveal the dynamics of how microbial agency can vary depending on the environment they are in, or in the presence of other microbial communities. These limitations led to many assumptions and misconceptions regarding the agencies of microbes that we now know to be incorrect. Pathogens were assumed to be more or less fixed entities in terms of their agency. There was also a belief that pathogens would gradually become less virulent over time (Méthot and Fantini, 2014, p. 4). Hinchliffe et al. (2016) describe this fixed, Euclidian imaginary of microbes as the basis for the antimicrobial warfare agenda between human and microbes.

Of relevance here too was the modern conception of what it was to be human at this time (or what was not human). These distinctions were influenced by the medical field that began to position microbes as biological outsiders to healthy bodies. This saw a shift in the way disease was approached – moving from a focus on the patient (inside) to a focus on external (outside) threats. This starts to inform a spatial dichotomy between a healthy inside and a pathogenic outside defining the modern concept of the human as an island, a biologically discrete or singular body. These models also positioned the notion of 'self' through the concept of the immune system, shaped in a way that is constantly under attack and which must secure itself from anything that is non-human or 'non-self'. In this way, the modern immune system is positioned as a barrier, protecting the clean bodily interior from the unclean external environment.

## ANTIBIOTIC ARCHITECTURE

While medical approaches focussed on the removal of the non-human from the human body, architecture followed suit with the agenda to remove the non-human from buildings and cities. It is here then I make an argument that antibiotic mindsets begin to permeate architectural thought. Distinctions between human and non-human, safe and unruly, and self and non-self led to design strategies that prioritised the separation of humans and non-human life. At the urban scale, animals including horses and livestock became increasingly translated out of the city while all unplanned nature was cleared and increasingly managed. Defensive notions of separation and boundaries begin to emerge in architectural language and strategies. Engineers began to specify hermetically sealed floors and walls, while new technologies began to permit mechanised control of indoor environments. Antiseptic products transitioned from the hospital to the home, informing antiseptic paints, and vacuum cleaners with microfilters emerged based on bacteriological science.

What we can also observe here is a shift in architectural mindset, from one that saw exposure to nature as necessary for health to one that favoured separation from nature as the basis for healthy building environments. This was exacerbated by emerging notions of climate and comfort. Le Corbusier's *mur neutralisant* (neutralising wall) was an early concept that favoured high airtightness to avoid air and heat flow between inside and outside to mechanically control the indoor climate (Gutiérrez, 2012). For example, the south-facing façade of the Cité de Refuge in Paris was completely sealed, with no opening windows. Such strategies of confinement were aligned with creating healthy, bioclimatic environments. Yet, when considered through the lens of indoor microbiome science, we can see how these strategies might also be seen to mark the start of indoor microbiomes that, over the following decades, became increasingly separated and disassociated from nature.

Architectural attention became increasingly focussed on issues of climate and well-tempered indoor environments (Banham, 1969), and the notion of separation here becomes exacerbated by technology, climate and contemporary environmental discourse. This shift towards separation would have had a measurable effect on preventing outdoor microbes becoming microbes in the indoor environment. The development of antibiotic medicines later in the mid-20th century was so successful in treating infectious disease that there was a collective confidence that the threat of infectious disease was over forever. Germ threats in buildings could now be managed with antiseptic chemicals, biocidal cleaning regimes and new technologies to control and purify the air, driven by a movement focussed on indoor air quality (Sundell, 2004).

**CONTEMPORARY PATHOLOGIES**
In the decades following the widespread use of antibiotic medicines, however, the confidence that infectious diseases were under human control began to erode. Emerging diseases such as AIDS, SARS and MERS bought a new germ panic, and this time microbes were revealing themselves to be far less fixed and linear than originally imagined. Microbes in the late 20th century became diagrammed as much more agile and threatening. They were now jumping between species and hijacking the increasingly connected world through new global transport links. Cases later in the century of the re-emergence of diseases that were thought to be previously defeated showed how microbes were evolving and mutating, threatening antibiotic security. This 'lively' understanding of microbes is most evident today through the emergence of AMR. In the introduction I presented AMR as an example of a contemporary pathology that has emerged over the last 50 years. Although it is a natural condition, its prominence today is understood as an example of 'antibiotic blowback' – a situation we have created from overly antibiotic strategies for managing life. Within the built environment, increasingly sterile environments that are dry and nutrient-poor are selecting for microbes and genes that are resistant to antibiotics. As these genes continue to evolve and spread, a report by the Review on AMR, chaired by Jim O'Neill, warns of potentially up to 10 million deaths globally per year by 2050 from AMR (O'Neill, 2016).

While the historical reading of architecture and health above might be summarised by health concerns traditionally associated with an abundance of microbes, today we also find ourselves faced with a series of new health challenges associated with a lack of microbes. These contemporary diseases have been described as 'epidemics of absence' (Velasquez-Manoff, 2012) or diseases associated with 'missing microbes' from bodies and environments (Blaser, 2014). Since the 1960s, while infectious disease has dropped, these new, immune-related diseases have risen sharply. These chronic pathologies are less deadly than infectious diseases; instead they are much slower, debilitating illnesses that severely diminish quality of life. Like the emergence of AMR, pathologies of absence can be seen as another example of antibiotic blowback. By creating ever more sterile environments, with buildings increasingly separated from nature and the outdoor environment, people born and living in cities are no longer being exposed to the microbial diversity required to sufficiently train normal immune function.

## TOWARDS A PROBIOTIC ARCHITECTURE

These contemporary pathologies present a tipping point. In the face of ongoing disease challenges, continuing the modern agenda of eradicating microbes from our lives will only serve to exacerbate these challenges further. In the following chapters I will return to these contemporary pathologies to inform new probiotic design approaches and show how probiotic science can be translated into new design practices that prioritise indoor microbiome restoration. Before that, in the second half of this chapter, I want to expand on the contemporary understanding of humans, microbes and their relationships that have emerged in recent years. Collectively, these positions challenge many of the modern concepts described above that underpin antibiotic mindsets. In doing so they make redundant the long-held assumption that healthy buildings should be free of microbes. I will outline these contemporary positions in order to then question how we might use them to inform new approaches to architecture that prioritise beneficial microbial exposure as a fundamental part of creating healthy and resilient cities.

## MICROBIOME SCIENCE

The development of new technologies in the late 20th century has radically shaped how we now study and understand microbial communities. These developments define the area of microbiome science, which is fundamental in underpinning the probiotic design approaches presented in this book. The first major development came in the 1970s following the discovery of DNA, which saw the emergence of polymerase chain reaction (PCR) and sequencing technologies. These genomic techniques facilitated the study of microbial communities in a way that was previously impossible using only traditional culturing techniques. The shift from the petri dish to sequencing techniques, and the subsequent use of the 16S rRNA gene for microbial community analysis, has been fundamental to facilitating the study of microbes within their environmental setting rather than in the laboratory.

Improved sequencing technologies at lower costs at the turn of the 21st century allowed for a vast accumulation of sequence data from studies such as the National Institutes of Health Human Microbiome Project (Turnbaugh et al., 2007) and the Earth Microbiome Project (Gilbert et al., 2018). These 'big data' approaches revealed the importance of microbiomes within higher organisms and made visible the critical roles that microbes play in human, non-human and planetary health. The imaginary of microbes which for so long was structured around their ability to disrupt has been completely transformed by microbiome science into one where microbes are now understood as fundamental to life and multispecies flourishing.

## THE HUMAN MICROBIOME

Central to this work then is the understanding of the human microbiome, a term which describes the multitudes of microbes that live inside and on the body. As well as revealing the sheer number of microbes that are present, it makes clear how the modern idea that a body should always be free of microbes is impossible. Not only would this be an impossibility, but it is also clear that the body would not function without them. Microbes are now understood to play key roles in metabolic, endocrinic and immunoregulatory functions, among others. Those in the gut are known to communicate with the brain, and are thought to influence not only perception but even our behaviour. Microbes here are understood not as non-human others to be protected from, but as fundamental and symbiotic parts of the human. This offers a crucial distinction between so-called good and bad microbes and, in doing so, challenges the modern linear model of all microbes as germs.

What is clear is that the development and maintenance of the human microbiome is fundamental for health. Acquisition of relevant microbes during early years is of particular importance in shaping immunoregulatory health. Parts of the microbiome are inherited – from the mother during birth and breastfeeding, but also from close contact with other family members. The rest, however, is acquired from the environment the body inhabits. This highlights how the environments we build and inhabit play crucial roles in shaping a human's microbiome, and subsequently its health through the types of microbes that exist in those spaces. These exposures are equally important throughout an organism's life. In line with evolutionary perspectives, these relevant microbes have been positioned less in relation to specific strains particularly, and more in line with the notion of microbial diversity that would be associated with natural environments containing diverse flora and fauna. These are outdoor, environmental microbes, those associated with soils, vegetation and farming – the types that the human body or gut would have evolved with over millions of years. Contemporary built environments and indoor microbiomes lacking in microbial diversity resulting from antibiotic mindsets appear not to offer these relevant exposures.

**THE HOLOBIONT**

The human microbiome offers a contemporary conception of the human that is radically different from the modern, discrete conception of the body described above. Instead, it is one described as a 'holobiont', a term first presented by Lynn Margulis (1991), where it referred to a biological entity involving a host and an inherited symbiont. It has since been extended to describe a host and its associated communities of microbes – the microbiome – and is now widely used and applied to humans, animals and plants. Contemporary questions of what it is to be human have been dominated by the holo-understanding of the microbiome, which challenges the modern definitions of individuality or self and non-self on which antibiotic biomedical models were formed (Tauber, 1994). Through its making visible of the millions of microbes that comprise the human body, the microbiome has seen *Homo sapiens* described instead as *Homo microbis* (Helmreich, 2014), where the body is cast as 10 per cent while the other 90 per cent is microbial (Sagan, 2013).

In the same way, the majority of our genes are inherited from microbes. Primarily, this challenges the biological distinction of human from non-human. This framework is now part of a major paradigm shift in biology that understands organisms less as genomically predicted from conception to organisms that are phenotypically determined through relationships between the host, its microbiome and its environment. This is radically different from the modern conception of the body where to be human was to be separated from nature. Instead, we are a multispecies body – a host/holobiont entanglement where the body is both one of the organism and a landscape for its microbial symbionts (Schneider, 2021).

**IMMUNITY AND HEALTH**

Finally, these contemporary positions also offer new conceptions of health and immunity that raise important questions regarding how we might seek to structure new relationships between our bodies and the building environment. The holobiont offers a more ecological conception of immunity where the defence of the self is replaced with one that is contextual and relational. Here immunity, and subsequently our identity, is constructed through dialogue with the organism's internal and external environments. Tauber (1994) suggests, in this view, that immune-determined identity should be understood instead through a balance between defence and ecological exchange. We might need to consider healthy built environments in the same way. A similar comparison might be made in line with the changing understanding of disease. Unlike the linear, modern model where disease was seen as a certain result of microbes crossing human boundaries, the medical field is beginning to shift from the germ theory of disease towards one that understands disease as more ecologically and socially relational (Hinchliffe et al., 2016). For these authors, disease is more than just microbes; instead it is a product of the relationships between the body, its microbes and its environment (social and physical). In line with this, an indoor microbiome might never be considered as distinctly healthy based on the composition of specific microbes. Instead it is one that is configurable, undergoing continuous exchanges that are shaped through the relationships between building design, occupants and microbes.

The importance of design in this model is central as the built and indoor environments are now the predominant habitats for humans in developed countries. Architectural design is known to affect the constitution of the indoor microbiome (Kembel et al., 2012, 2014; Meadow et al., 2014a), and therefore the microbiome of a building is implicit in structuring the microbiome of the human (and vice versa) – the two cannot be distinguished as they are fundamentally related. These authors suggest that, rather than focussing on practices of presence or exclusion, the focus on health should look towards the configuration of the quality and spatial relations of these three factors. The concept of the microbiome then, I argue, is of fundamental relevance to rethink the architectural design of healthy buildings. Understanding the role of design in shaping the indoor microbiome is therefore important if we are to explore the potential to begin to inform the relationships that can structure a healthy indoor microbiome.

## MICROBIOMES OF THE BUILT ENVIRONMENT

This brings us then to the microbiomes of the built environment. Terminology varies in the literature, but includes 'indoor microbiome', 'building microbiome' and 'microbiome of the built environment'. Like living organisms, buildings have their own microbiomes too. When viewed through the notion of the microbiome, buildings can be considered as vast landscapes of different habitats and niches at multiple scales that are colonised by multiple, diverse groups of microorganisms (Adams et al., 2016; Gilbert & Stephens, 2018). As well as colonisation, both the material fabric and the environmental ether of a building provide conditions for microbial succession and transmission to humans (Li et al., 2021). Microbes in buildings are found in three sources – in the air, on surfaces or in water systems. The types and constitution of these microbes include those living in the building, those that are brought in from the outside (environmental microbes) and those that come from occupants (human-associated microbes).

In the same way that humans have microbiomes that are unique to the individual, the literature shows that the constitution of a building's indoor microbiome can be considered equally unique. This is influenced by external factors such as location, climate, surrounding geography (Adams et al., 2016; Chase et al., 2016; Gibbons, 2016) and the building's occupants (humans and pets). Neither are building microbiomes fixed: they comprise diverse communities of bacteria, viruses and eukaryotes that appear to be continually changing and interacting in different ways. Some microorganisms can be actively growing, while some may be dormant, surviving using various mechanisms until moisture becomes available. Some may be dying or dead. These are still of relevance for health.

The term microbiome has multiple definitions in the literature, depending on the category and context of study (Berg et al., 2020). In this book, I use a combined definition offered by Ho and Bunyavanich (2018), which describes the microbiome as 'the sum of the microbes and their genomic elements in a particular environment'. Here, the term microbiome is used distinctly from the term 'microbiota', which typically refers only to living organisms in a defined environment (Marchesi & Ravel, 2015). This would usually discount viruses, phages and free or extracellular DNA, which are considered non-living (Dupré & O'Malley, 2009). Instead, microbiome incorporates these as well as other 'mobile genetic elements', and also considers the agencies of many molecules and metabolites produced by microorganisms which are structured in line with the surrounding environment (Berg et al., 2020). These non-living microbial components are of importance to this research as they are also known to have both positive and negative effects on occupant health. Some secondary metabolites produced by bacterial and fungal microbiota in damp homes are known to contribute to health problems associated with sick building syndrome (Andersson et al., 1998), while others are known to stimulate immunoregulatory proteins that help protect against allergies (Schuijs et al., 2015).

**BUILDING DESIGN**

Building design impacts the constitution of the indoor microbiome, which is seen to differ considerably across different building typologies (Kembel et al., 2012, 2014). The literature shows measurable differences across homes, hospitals, offices, transport systems and even space stations; and it is now generally understood that the typology of a building will impact its microbiome through its design, its associated functions and its level of occupancy (Adams et al., 2016). High occupancy is associated with high measures of human-associated microbes (Kembel et al., 2012; Meadow et al., 2014b). The manner in which a building is occupied in relation to typology will also determine the biological taxa that are present (Dannemiller et al., 2016). For example, a home or dwelling that typically contains the same, small number of people presents a very different microbial condition compared to a public building that contains lots of people and is constantly in flux. It is also clear that a building's microbiome is time-based and will vary over time (Rintala et al., 2008). The constitution of these communities is continually combining and shifting as microbes enter or leave a building as a result of changing environmental conditions, seasonal changes or even changes in the building's function or occupancy.

The home condition has been the subject of much study in this area, and source-tracking techniques used by Lax et al. (2014) have shown that a person travelling away from home for a short period results in the rapid decline of the bacterial communities associated with that person, but then a rapid increase on home surfaces when the person returns. Results from the Home Microbiome Project show how, after a family moves into a new house, the microbiome of that house rapidly converges towards the communities found in the new occupants' former house (Gilbert, 2014). A similar pattern was seen in schools, where a study found that human occupancy in classrooms increased the bacterial genome concentration of indoor air by nearly two orders of magnitude compared to holiday periods when the rooms were unoccupied (Hospodsky et al., 2012).

## BUILDING ATTRIBUTES

Alongside external factors and occupants, it is evident that, within a building's typology, specific attributes and design strategies will play roles in shaping the indoor microbiome. Strategies relating to sunlight, building materials, and heating, ventilation and air conditioning (HVAC) have all been identified as relevant through the resulting environmental conditions they create, including temperature, humidity and airflow. However, understanding specifically how building function and design impact the indoor microbiome remains a knowledge gap in the literature. Challenges include the interrelatedness of these attributes and what this means for the microbial niches that result. While more research is needed, it is clear that, for buildings to be understood in a way that we might seek to inform the microbial and ecological conditions, building attributes cannot be considered independently. For example, the building's typology will influence material choice, and fenestration strategies will in turn affect the environmental conditions in relation to the HVAC system. These factors will also inform the connectedness of the building's spaces, circulation and other occupant activities, such as density, movement and behaviours. It is the configuration and culmination of these factors that will determine colonisation, succession and potential for transmission.

**CONCLUSION**

In order to develop an understanding of what a probiotic approach to architecture might involve, in this chapter I have sited the agenda of the work within the historical dialogue between architecture, health and disease. We have read how over-management of life using antibiotic strategies in line with the germ theory of disease, combined with a shift towards increasingly sealed buildings in the second half of the 20th century, has shaped indoor microbiomes that are now pathological through a lack of microbial diversity. In line with a contemporary understanding of human–microbe relationships, there is a need for broader, more holistic approaches to designing buildings in relation to context of the Anthropocene, sustainability and equality that must also involve microbes and human health. Just as the understanding of the holobiont body rejects inside/outside biological distinctions, architecture must also reject the principle of separation as its predominant mode of biopower.

It is clear that design plays a key role in shaping the constitution of microbes in buildings; and so, by employing strategies that prioritise microbial diversity, it may be possible to rewild buildings with missing microbes that are fundamental to shaping holobiont health. In practice, this offers a mix of both direct and indirect strategies across a range of scales that might be employed to design built environments in consideration of the microbiome as the basis for health. If our current cities are lacking sufficient microbial diversity, new design approaches are needed that can increase biodiversity at the urban scale. Importantly however, we must also facilitate the transfer of these environmental microbes into buildings and radically recalibrate our indoor environments to support microbial presence. In some cases, we may need new ways to directly reintroduce microbes in order to sufficiently rewild buildings.

I will use the rest of this book to explore these approaches, which raises more practical questions, such as: Which microbes should be in a building? Where in the building should they be? And how would we put them there? These questions will be addressed in the coming chapters. The fields of biodesign and living architecture will be implicit in shaping this agenda where, alongside microbes as material or energy producers, they might also be seen as symbiotic partners, essential for holobiont health. We need to develop new methodologies for design, but also ways of testing and monitoring these approaches. It will be crucial to understand how the use of living materials and species as a fundamental part of buildings will shape the microbiomes of the buildings and the humans they are designed for.

**REFERENCES**

Aalto, A., Frampton, K., Korvenmaa, P., Pallasmaa, J., & Treib, M. (1998). *Alvar Aalto: between humanism and materialism*. Abrams.

Adams, R. I., Bhangar, S., Dannemiller, K. C., Eisen, J. A., Fierer, N., Gilbert, J. A., Green, J. L., Marr, L. C., Miller, S. L., Siegel, J. A., Stephens, B., Waring, M. S., & Bibby, K. (2016). Ten questions concerning the microbiomes of buildings. *Building and Environment*, *109*, 224–234. https://doi.org/10.1016/j.buildenv.2016.09.001.

Andersson, M. A., Mikkola, R., Kroppenstedt, R. M., Rainey, F. A., Peltola, J., Helin, J., Sivonen, K., & Salkinoja-Salonen, M. S. (1998). The mitochondrial toxin produced by *Streptomyces griseus* strains isolated from an indoor environment is valinomycin. *Applied and Environmental Microbiology*, *64*(12), 4767–4773.

Armstrong, G. L., Conn, L. A., & Pinner, R. W. (1999). Trends in infectious disease mortality in the United States during the 20th century. *Jama*, *281*(1), 61–66.

Banham, R. (1969). *The architecture of the well-tempered environment*. Architectural Press.

Berg, G., Rybakova, D., Fischer, D., Cernava, T., Vergès, M.-C. C., Charles, T., Chen, X., Cocolin, L., Eversole, K., Corral, G. H., Kazou, M., Kinkel, L., Lange, L., Lima, N., Loy, A., Macklin, J. A., Maguin, E., Mauchline, T., McClure, R., ... & Schloter, M. (2020). Correction to: Microbiome definition re-visited: old concepts and new challenges. *Microbiome*, *8*(1), 1–22. https://doi.org/10.1186/s40168-020-00905-x.

Blaser, M. J. (2014). *Missing microbes : how killing bacteria creates modern plagues*. Simon & Schuster.

Carter, S. (2007). *Rise and shine: sunlight, technology and health*. Berg.

Chase, J, Fouquier, J., Zare, M., Sonderegger, D. L., Knight, R., Kelley, S. T., Siegel, J, & Caporaso, J. G. (2016). Geography and location are the primary drivers of office microbiome composition. *mSystems*, *1*(2):e00022-16. https://doi.org/10.1128/mSystems.00022-16.

Dannemiller, K. C., Gent, J. F., Leaderer, B. P., & Peccia, J. (2016). Influence of housing characteristics on bacterial and fungal communities in homes of asthmatic children. *Indoor Air*, *26*(2), 179–192.

Drake, S. (1997). The architectural antimephitic: modernism and deodorisation. *Architectural Theory Review*, *2*(2), 17–28.

Dupré, J., & O'Malley, M. A. (2009). Varieties of living things: life at the intersection of lineage and metabolism. *Philosophy and Theory in Biology*, *1*, e003. doi:10.3998/ptb.6959004.0001.003.

Gibbons, S. M. (2016). The built environment is a microbial wasteland. *mSystems*, *1*(2), e00033-16. https://doi.org/10.1128/mSystems.00033-16.

Gilbert, J. A. (2014). *The Home Microbiome Project.* Argonne National Lab (ANL), Argonne, IL.

Gilbert, J. A., & Stephens, B. (2018). Microbiology of the built environment. *Nature Reviews Microbiology, 16*(11), 661–670.

Gilbert, J. A., Jansson, J. K., & Knight, R. (2018). Earth microbiome project and global systems biology. *mSystems, 3*(3), e00217-17.

Gutiérrez, R. U. (2012). 'Pierre, revoir tout le système fenêtres': Le Corbusier and the development of glazing and air-conditioning technology with the Mur Neutralisant (1928–1933). *Construction History, 27,* 107–128.

Helmreich, S. (2014). *Homo microbis:* the human microbiome, figural, literal, political. *Thresholds, 42,* 52–59.

Hinchliffe, S., Bingham, N., Allen, J., & Carter, S. (2016). *Pathological lives: disease, space and biopolitics.* John Wiley & Sons.

Ho, H., & Bunyavanich, S. (2018). Role of the microbiome in food allergy. *Current Allergy and Asthma Reports, 18*(4), Article 27. https://doi.org/10.1007/s11882-018-0780-z.

Hobday, R. A., & Dancer, S. J. (2013). Roles of sunlight and natural ventilation for controlling infection: historical and current perspectives. *Journal of Hospital Infection, 84*(4), 271–282.

Hospodsky, D., Qian, J., Nazaroff, W. W., Yamamoto, N., Bibby, K., Rismani-Yazdi, H., & Peccia, J. (2012). Human occupancy as a source of indoor airborne bacteria. *PLOS ONE, 7*(4), e34867.

Kembel, S. W., Jones, E., Kline, J., Northcutt, D., Stenson, J., Womack, A. M., Bohannan, B. J. M., Brown, G. Z., & Green, J. L. (2012). Architectural design influences the diversity and structure of the built environment microbiome. *ISME Journal, 6*(8), 1469–1479.

Kembel, S. W., Meadow, J. F., O'Connor, T. K., Mhuireach, G., Northcutt, D., Kline, J., Moriyama, M., Brown, G. Z., Bohannan, B. J. M., & Green, J. L. (2014). Architectural design drives the biogeography of indoor bacterial communities. *PLOS ONE, 9*(1), e87093.

Lax, S., Smith, D. P., Hampton-Marcell, J., Owens, S. M., Handley, K. M., Scott, N. M., Gibbons, S. M., Larsen, P., Shogan, B. D., Weiss, S., Metcalf, J. L., Ursell, L. K., Vázquez-Baeza, Y., van Treuren, W., Hasan, N. A., Gibson, M. K., Colwell, R., Dantas, G., Knight, R., & Gilbert, J. A. (2014). Longitudinal analysis of microbial interaction between humans and the indoor environment. *Science, 345*(6200), 1048–1052. https://doi.org/10.1126/science.1254529.

Li, S., Yang, Z., Hu, D., Cao, L., & He, Q. (2021). Understanding building–occupant–microbiome interactions toward healthy built environments: a review. *Frontiers of Environmental Science & Engineering, 15*(4), 1–18.

Magner, L. N. (2002). *A history of the life sciences, revised and expanded.* CRC Press.

Marchesi, J. R., & Ravel, J. (2015). The vocabulary of microbiome research: a proposal. *Microbiome, 3,* Article 31. https://doi.org/10.1186/s40168-015-0094-5.

Margulis L. (1991). Symbiogenesis and symbionticism. In L. Margulis & R. Fester (eds), *Symbiosis as a source of evolutionary innovation: speciation and morphogenesis* (pp. 1–14). MIT Press.

Meadow, J. F., Altrichter, A. E., Kembel, S. W., Kline, J., Mhuireach, G., Moriyama, M., Northcutt, D., O'Connor, T. K., Womack, A. M., Brown, G. Z., Green, J. L., & Bohannan, B. J. M. (2014a). Indoor airborne bacterial communities are influenced by ventilation, occupancy, and outdoor air source. *Indoor Air, 24*(1), 41–48. https://doi.org/10.1111/ina.12047.

Meadow, J. F., Altrichter, A. E., Kembel, S. W., Moriyama, M., O'Connor, T. K., Womack, A. M., Brown, G. Z., Green, J. L., & Bohannan, B. J. M. (2014b). Bacterial communities on classroom surfaces vary with human contact. *Microbiome, 2*(1), 1–7.

Méthot, P. O., & Fantini, B. (2014). Medicine and ecology: historical and critical perspectives on the concept of 'emerging disease'. *Archives internationales d'histoire des sciences, 64*(172–173), 213–230.

Muller, N., Werner, P., & Kelcey, J. G. (2010). *Urban biodiversity and design.* John Wiley & Sons.

O'Neill, J. (2016). *Tackling drug-resistant infections globally: final report and recommendations.* Review on Antimicrobial Resistance. https://amr-review.org/sites/default/files/160518_Final%20paper_with%20cover.pdf.

Ozonoff, V. V. (1982). *A healthy and salubrious place: public health and city form.* MS thesis, Massachusetts Institute of Technology.

Rintala, H., Pitkäranta, M., Toivola, M., Paulin, L., & Nevalainen, A. (2008). Diversity and seasonal dynamics of bacterial community in indoor environment. *BMC Microbiology, 8*(1), 56. https://doi.org/10.1186/1471-2180-8-56.

Rook, G. A. W., & Bloomfield, S. F. (2021). Microbial exposures that establish immunoregulation are compatible with targeted hygiene. *Journal of Allergy and Clinical Immunology, 148*(1), 33–39.

Sagan, D. (2013). *Cosmic apprentice: dispatches from the edges of science.* University of Minnesota Press.

Schneider, T. (2021). The holobiont self: understanding immunity in context. *History and Philosophy of the Life Sciences, 43*(3), 1–23.

Schuijs, M. J., Willart, M. A., Vergote, K., Gras, D., Deswarte, K., Ege, M. J., Madeira, F. B., Beyaert, R., van Loo, G., & Bracher, F. (2015). Farm dust and endotoxin protect against allergy through A20 induction in lung epithelial cells. *Science, 349*(6252), 1106–1110.

Sundell, J. (2004). On the history of indoor air quality and health. *Indoor Air, 14*(s7), 51–58.

Tauber, A. I. (1994). The immune self: theory or metaphor? *Immunology Today, 15*(3), 134–136.

Tomes, N. (2000). The making of a germ panic, then and now. *American Journal of Public Health, 90*(2), 191–198.

Turnbaugh, P. J., Ley, R. E., Hamady, M., Fraser-Liggett, C. M., Knight, R., & Gordon, J. I. (2007). The Human Microbiome Project. *Nature*, *449*(7164), 804–810.

Velasquez-Manoff, M. (2012). *An epidemic of absence: a new way of understanding allergies and autoimmune diseases*. Simon & Schuster.

Wigley, M. (2001). *White walls, designer dresses: the fashioning of modern architecture*. MIT Press.

# Micro-Scale: Probiotic Materiality

## Chapter 3

DOI: 10.4324/9781003207917-3

## MICROBIAL INTERVENTIONS IN THE BUILT ENVIRONMENT

In Chapter 2, I presented a broad overview of both the scientific and philosophical basis for a necessary recalibration of architectural approaches towards health that must align with contemporary understanding of the symbiotic relationships between humans and microbes. To design in this way requires a shift from antibiotic mindsets to probiotic mindsets which can underpin how we might facilitate and manage non-human life in the built environment in a way that permits multispecies flourishing at multiple scales. I also introduced the relationship between building design and the resulting microbes that shape the indoor microbiome. These building–microbe relationships raise the possibility that human–microbial exposure might be modulated through architectural design interventions that improve human health by creating exposure to beneficial microbes. It is to these interventions and how we might design them that we now turn.

This and the following two chapters will outline the development of this probiotic design approach operating at multiple scales – from the micro-scale of the material matrix in this chapter to the meso-scale of the indoor environment in Chapter 4 and up to the macro-scale of the city in Chapter 5. These probiotic design approaches and interventions will be based on the key aim of continuing to limit exposure to harmful microbes, but, crucially, will also permit and facilitate exposure to beneficial microbes.

While many strategies exist to limit harmful microbes, most are antibiotic in their approach and very few exist to promote beneficial exposure. Those that do are mostly 'indirect' approaches that aim to increase the transport of outdoor microbes inside. The work here will focus specifically on interventions that seek to 'directly' add beneficial microbes to buildings. Building on the bioreceptive design paradigm, I explore the potential to directly inoculate building materials, surfaces and spaces with benign microbes that are beneficial for health. These approaches will be described through a body of experimental research undertaken over the last five years. The work seeks, in its broadest sense, to challenge the modern assumption that 'healthy spaces = fewer microbes'. If the direct challenge to that statement is to replace it with one that says 'healthy spaces = more microbes', questions begin to arise in relation to *which* microbes we should put into buildings and *how* might we do this.

As a starting point, I return to the contemporary pathologies introduced in Chapter 2. These are the new microbial diseases that have emerged over the last 50 years that show the modern city as pathological through its lack of biodiversity at multiple scales. In this chapter, I will stay away from diseases of missing microbes. Instead, I will begin by discussing the potential for a probiotic design approach to inform healthier buildings in line with the contemporary pathology of antimicrobial resistance (AMR). While AMR is a broad challenge across a range of fields, architecture and the way we design and plan buildings in relation to microbes are implicit in its emergence and spread in the built environment (Mahnert et al., 2019). In rejecting antibiotic approaches that eradicate life, the work here explores instead the potential for using life to manage life. The hypothesis is that we can utilise mechanisms of beneficial microbes to inhibit pathogenic microbes rather than using antibiotic materials or chemicals. This approach uses good bacteria to reduce the presence and persistence of pathogenic microbes; but, crucially, it does so in a manner that does not exacerbate microbial evolution towards more resistant states, while still permitting the presence of other benign and beneficial microbes as a healthy part of indoor environments. In this way, the question of which microbes might be directly added to indoor environments can be addressed through the selection of benign strains that can limit the presence of pathogens in buildings.

## NOTBAD

The work began initially with 'Niches for Organic Territories in Bio-Augmented Design' (NOTBAD), a project funded by the UK's Arts and Humanities Research Council (AHRC) to explore how design research could develop novel ways to limit the spread of antimicrobial resistant organisms (AMR) in buildings. AMR organisms comprise the so-called 'superbugs' that have evolved resistance to modern antibiotic drugs and now pose a renewed threat to human life from simple infection. At first glance, the emergence and persistence of AMR seems to fit the familiar narrative of bacterial pathogens waiting to attack, outsmarting us by evolving quicker than we can respond. It is often positioned publicly as microbes 'fighting back', in line with the familiar warfare narrative between humans and microbes described in Chapter 2. As we have read, however, emerging evidence shows how overuse of and overreliance on antibiotic approaches are in fact part of the problem. Rather than microbes fighting back, it seems that, by attempting to eradicate microbes from our bodies and our environments, we are creating conditions that are causing microbes to evolve resistance.

The acronym for the project was a play on the contemporary understanding of microbes that differentiates between so-called good and bad microbes. Instead of using antibiotic materials and chemicals, we explored the potential for integrating beneficial strains of bacteria into materials that could then out-compete bad or 'pathogenic' strains of bacteria. Crucially, this approach would allow other benign microbes to remain, and in a manner that could avoid creating conditions that would select for resistance. It became the first example of what I define as a 'probiotic' design approach to architecture, which uses bacteria to create healthy building environments rather than excluding them. It reconceptualises materials from those that are healthy through their sterility to those that are healthy through the microbes they support.

## BUILDING SURFACES AS MICROBIAL ECOLOGIES
Materials in this sense need to be understood through their ability to serve as a niche for microbial colonisation. So what do we know from microbiome research about building materials and microbes? Microbes can be observed on all building materials and surfaces (Li et al., 2021), even on those designed to be actively antimicrobial (Hu et al., 2019). However, despite the revelations from DNA sequencing studies that have painted a picture of microbes in buildings as everywhere, on all surfaces around us, the idea that these microbes are thriving and growing akin to how they do in a petri dish appears not to be the case. Surfaces in buildings have been generally described as 'microbial wastelands' (Gibbons, 2016) and compared to desert conditions where nutrients and moisture are scarce. Microbes are, more often than not, mostly 'surviving' on the majority of building surfaces.

Generally, in the contemporary city most building surfaces are designed and applied so as to reduce moisture and nutrient availability, both through their material specification and their physical properties. Predominantly, these strategies are employed to limit microbial presence and proliferation, and some indoor surfaces are designed specifically to be antimicrobial. Examples include micro-scale patterning on high-touch surfaces such as door handles to reduce microbial fouling (Vasudevan et al., 2014) or paints impregnated with antimicrobial coatings. Many other materials and finishes that are not directly antimicrobial still seek to limit microbial presence. Glazing, polishing, sealing, plastering and painting offer aesthetics of newness and cleanliness but are still underpinned by strategies that seek to minimise moisture retention and dust collection and facilitate easy cleaning/wiping.

These material strategies shape microbiomes that are low in microbial load; yet they also limit microbial diversity, selecting for specific microbes that are able to survive in these conditions. Some bacteria can sporulate under these stress conditions, changing their phenotype into low-energy, dormant states in which they can survive for extremely long periods of time without moisture or nutrients. In this state, they are able to regerminate and potentially start growing if moisture does become available (Gibbons et al., 2015; Hegarty et al., 2018). These microbiomes are heavily shaped towards ones that select for survival; and, more worryingly, they appear to be selecting for antimicrobial resistance.

## MAN-MADE AMR

Specifically, this describes what are called 'man-made microbial resistances' in built environments. By creating strong selective pressure, especially on building surfaces, buildings are facilitating evolutionary mechanisms towards drug-resistant states (Fahimipour et al., 2018). Microbiome studies demonstrate how overly managed buildings drive microbiomes to encode for genes associated with oxidative stress that promote survival (Mahnert et al., 2019). This leads to increased functional mechanisms such as membrane transport and secretion, mechanisms which microbes use to survive when moisture and nutrients are scarce.

Even more challenging is that AMR has spread more quickly and more vigorously than initially predicted. This is partly a result of horizontal gene transfer – a mechanism whereby genes can be laterally acquired from other microbes in close proximity – as well through the more understood hereditary or vertical process. Other, opportunistic microbes are able to acquire these genes and functional mechanisms through this process. Antibiotically managed surfaces are observed to induce direct and cross-resistance to antibiotics (Kampf, 2018) and overall are shaping microbiomes towards higher levels of virulence, defence and resistance (Mahnert et al., 2019). These mechanisms are advantageous for the microbes, but present significant health risks for humans.

While there is a clear need for some parts of certain buildings to be free of microbes, for example surgical spaces in hospitals, there are many others (such as homes and public buildings) that do not need to be. Antibiotic cleaning products also expose children to agents that are believed to exacerbate allergic responses to normally innocuous antigens (Rook & Bloomfield, 2021). This suggests that sterile environments for human habitation should be limited to those strictly necessary, and that other non-specific areas of built environments should be designed and managed in ways that are hygienic but crucially, not sterile.

We can make the case then that healthy surfaces should also permit the presence of other microbes. A degree of microbial load on surfaces, especially one which constitutes a degree of diversity of microbial strains, is now understood to offer microbial 'competition'. Indiscriminate removal of all microbes from surfaces also removes microbial competition for space and nutrients, resulting in uncontested breeding grounds for pathogens with resistance to proliferate. Microbial diversity therefore plays an important role in stabilising microbial communities on surfaces, acting as a kind of shield against pathogen proliferation (Blaser, 2016; Kennedy et al., 2002).

## RECONSIDERING BUILDING MATERIALS AS BENEFICIAL MICROBIAL NICHES

The bioreceptive design paradigm we developed as part of the BiotA Lab sought to reconceptualise the building fabric from one that is protective but inert to one which could protect but also serve as a host for a range of biological systems (Cruz & Beckett, 2016). Where these approaches had focussed previously on external building façades and outdoor environments, knowledge of material bioreceptivity came from the fields of building heritage and preservation. Here, the literature had predominantly addressed the negative associations of biological growth on materials due to their potential to damage historic buildings and structures of importance. We inverted these concepts to inform how we could utilise these negative design parameters to instead encourage the growth of microorganisms on building materials in line with a climatic agenda but in a way that didn't damage the material. A similar approach can be considered for indoor materials, where their condition can be reconceptualised from surfaces that seek to minimise microbial presence to surfaces that provides niches for beneficial microbes to inhabit. To explore this, it is necessary to understand the parameters that shape the material–microbe interface in indoor building environments.

Indoor microbiome knowledge relating to microbial growth and presence indoors has mostly focussed on the negative associations of microbial presence in buildings. Typically this is discussed through conditions of building failure, either in relation to moisture and damp or to microbial transmission of pathogens. A probiotic approach questions how we might use this knowledge in a positive manner to facilitate microbial colonisation in ways that are beneficial for health. The term 'bioreceptivity' is little used in these fields, though many of the parameters are similar. The physical and chemical properties of a material, including pH level and moisture-absorbing potential, are understood to determine the specific ability of that material to support growth. Smooth, impervious, easily cleanable materials and geometries have become the default in the mindset of microbial security. Instead, materials designed to exhibit porosity, surface roughness and textural geometries can serve as microbial niches for specific conditions.

Material porosity and surface roughness are understood as key parameters that can support microbial colonisation due to their ability to trap cells, along with dust and other organic compounds, and to retain moisture (Lax et al., 2019). Porous materials could be inoculated with beneficial microbial communities and introduced into buildings as surfaces or components. This methodology has proved successful in relation to the growth of photosynthetic strains of algae and bryophytes, promoting confidence that similar approaches could be developed to support strains of beneficial bacteria once the conditions needed for their survival and growth could be determined.

## MICROBIAL IDENTIFICATION

We return now to the question of *which* microbes might be beneficial. In *The Probiotic Planet*, Lorimer (2020) identifies a series of key practices that show how probiotic science is being translated into rewilding and biome restoration approaches. An important first step in landscape rewilding involves the 'identification' of keystone species that are selected for their ability to engineer the ecosystems in which they will operate. Identification of bacterial strains to use is a key first step in order to lead the material development. The specific AMR focus of the NOTBAD project offered a starting point from which to address this question and narrowed this search to consider strains of bacteria that, firstly, are known to be benign to humans and, secondly, that exhibit some known inhibitory mechanism against AMR pathogens. The target pathogen in this case was *Staphylococcus aureus*, better known in its resistant form as MRSA, the hospital superbug which is now resistant to several widely used antibiotics.

We began then by exploring strains of *Bacillus subtilis*, a soil-derived, Gram-positive, environmental bacterium well known to be benign to humans, and which exhibits a range of functions and agencies that are potentially beneficial to humans and the built environment. Specific strains were identified based on their known ability to produce antimicrobial mechanisms along with other molecules that prevent the adhesion and accumulation of other microorganisms on surfaces (Duc et al., 2004). Important too for this identification stage was to consider strains that are likely to survive in indoor environments where nutrients and moisture may be limited. *B. subtilis* strains are known to form spores that can remain dormant for hundreds of years, staying highly resistant to alkaline pH, heat and other environmental stresses such as dehydration (Setlow, 2014).

**MATERIAL SELECTION**

The next stage sought to develop materials that could provide a specific bioreceptive niche for these strains based on their specific requirements. Here, we decided to explore the use of materials that are already common in architecture but that can exhibit porosity as a result of the fabrication process. Here, 'manually mixed materials' such as concretes and ceramics were explored, whereby variations in aggregate size, water/binder ratios and compaction can result in targeted pore sizes. We sought to avoid expensive materials that rely on costly, pharmaceutical-grade materials, such as those that can only be made in small quantities or whose scaling up would be limited by either available technology or cost. Hence, we aimed to develop materials that are potentially feasible for real-world use in the built environment.

Based on these criteria, various materials were selected for initial exploration, as shown in Figure 3.1. These included: ceramics (common in buildings, typically used for wall/floor tiles and other objects); concretes (also common in buildings, typically structural but also used non-structurally on floors, walls and panels); 3D-printed materials including plastic (Nylon PA2200 via Selective Laser Sintering/SLS technology, potentially to be used for components and objects); and 3D-printed aluminium (via Direct Metal Laser Sintering/DMLS technology, potentially for components, frames and door handles). Through this process, we aimed to define a methodology of dosage and fabrication to produce material samples according to target values of pH and porosity for *B. subtilis* strains.

Figure 3.1 A range of materials tested for material–microbe viability

**PROBIOTIC METHODOLOGIES**

We also needed to develop a methodology for integrating the bacterial strains into the material matrix, and then for assessing the biocompatibility of the material/microbe hybrids. Previous bioreceptive design work used the notion of visible biological growth as the measurable factor in which to test for successful integration. In the case of probiotic materials containing bacterial cells, it was evident that these microbes might never be visible to the eye in the same way as bryophytes and other cryptogams are. It was also clear early on in the work that the notion of 'growing' bacteria in materials might not be the relevant term to use. Instead, the work focussed on whether the bacterial cells could 'survive' and remain 'viable' within the material in order to demonstrate their beneficial agency over time. To do this, we would need to develop methodologies to measure whether, following inoculation into the material matrix, and beyond, they would be able to survive over time without the addition of moisture or nutrients.

The final step, if these first two stages were successful, was to develop a methodology to test their probiotic agency to inhibit MRSA. This methodology is shown in Figure 3.2 and was defined in three stages: 1 – Material/Microbe Viability, an initial process to identify and select materials suitable for bacterial integration; 2 – Microbial Survival, assessment of the long-term survival of bacteria in these materials; 3 – Pathogen Inhibition, a final stage where selected hybrid materials could be tested directly against a known pathogen to validate its probiotic action.

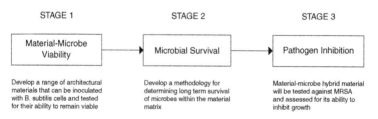

Figure 3.2 Experimental methodology for the design of probiotic material

**Stage 1: Material/Microbe Viability**

Operating at the micro-scale of the material–bacteria interface, materials here are considered and designed according to the requirements of the microbes. This defines a microbial-led process while still trying to ensure that the materials remain viable for architectural-scale objects in terms of cost and strength. Material samples were produced in the studio, where their physical and chemical properties were targeted, characterised and tweaked according to literature values for the chosen strains or communities of bacteria. These were then taken to the laboratory, where microbial cells were cultured and inoculated into the material volume and microbiological methodologies used in order to determine success, failure and iteration (as shown in Figure 3.3).

We approached this initial exploration phase in a manner common to other biodesign projects that have engaged a DIY material approach in line with what has been described as a kind of material 'tinkering' stage (Karana et al., 2018). This approach leans towards materials science methods, using specific weights and material mixes, but seeks to assess them in a way that avoids the slow and

Figure 3.3 Experimental methodology for determining material–microbe viability

expensive nature of rigorous microbiological characterisation at this early stage. Material porosity was defined as a key property, firstly to permit inoculation of cells in a liquid culture, but also to provide a physical scaffold of micro-enclosures or niches within the material volume where bacteria can physically attach.

The scale of the work here was driven mostly by the microbiological methodologies. We were working at the scale of 1 $cm^3$ cubes, which were cheap and easy to make in multiple repeatable samples as required. They were also a suitable size for the associated microbiological equipment, meaning that, at this stage of the work, we did not need to make large quantities of the microbial inoculant.

The mixed materials (including concretes and ceramics) were fabricated using a dry-mix methodology where aggregates of known particle sizes were mixed with low binder ratios to ensure interstitial pores remained once cured or bonded by glass during firing. These were cast and compressed in moulds to form samples. A computational method was developed for the 3D printed metal and plastic materials, designed computationally as volumes filled with triangular lattices with designated thicknesses resulting in sub-millimetre pore sizes between lattices. Figure 3.4 shows scanning electron microscopy (SEM) images of the various material porosities achieved using these fabrication approaches. Individual particle aggregates are visible creating interstitial pore spaces between them.

In order to assess these material mixes, feedback and success at this point were determined according to the biological notion of microbial viability. This primarily sought to determine whether the bacteria were able to survive once embedded into the material, defined as 'material–microbe viability'. This relates initially to a short-term notion of viability to ensure the microbes can survive the inoculation step into the material matrix. Following inoculation into the material, the samples were stored at ambient conditions for one week and then replated on to nutrient plates. Any observable growth on the nutrient plate could be taken as indicative that the microbes could remain viable within the material matrix.

Material samples were then taken to the laboratory for inoculation with *B. subtilis* cells. Samples were sterilised using an autoclave and stored under sterile conditions prior to inoculation.

Figure 3.4 SEM images of casting material samples showing porosity, from left to right: ceramic, SLS nylon and concrete

The *B. subtilis* strain was then grown in a liquid culture to a known optical density whereby 300 μL were hand pipetted into the material samples. Inoculated samples were then stored under ambient indoor conditions. Over the following days, samples were periodically replated to nutrient agar plates to assess material/microbial viability, as evidenced by observable growth on the plate (Figure 3.5).

This quick and easy methodology served as an early screening method for the initial range of materials. We saw that the majority of materials appeared to demonstrate some level of material microbe viability. This aligns with the literature understanding of the indoor microbiome which seems to suggest that microbes are able to colonise most materials in buildings. What appears to be the case is that material porosity is important for providing physical niches of protection for microbial cells to colonise.

Figure 3.5 Testing material–microbial viability through regermination

There were some material mixes that did not demonstrate microbial viability. Figure 3.5 shows two concrete mixes, one made using a magnesium phosphate cement (MPC) binder and the other using CEM III/B cement. The MPC samples did not demonstrate regermination, suggesting that the microbes were not able to survive in the material matrix. Within the context of the project, we were not able to look further into reasons for this. Instead, we used this stage as a way of narrowing down the material options towards materials that offered the best chance of successful bio-integration.

### Stage 2: Microbial Survival

In this second stage of the methodology, we sought to better understand longer-term notions of microbial survival in buildings. Although the work had no defined time frame, there was a sense that survival over the course of months would be necessary for use in buildings. If survival times were shorter than this, materials would require regular reapplication, bringing associated costs through the need for maintenance. Time and costs associated with high maintenance needs have proven problematic for other living systems used in architecture, such as plants and green walls (Perini & Rosasco, 2013). A major challenge for larger living systems and their application to architecture has been the need for regular supplies of nutrients and water for survival. Therefore a key aim of the present research was to avoid expensive or intensive maintenance systems. It may not be necessary, however, for beneficial microbes in buildings to be kept in a state of constant germination.

An important mechanism of microbes is their ability to exist in different states. When nutrients and moisture are available, cells exhibit a germinated state; but when these resources are scarce, some strains are able to sporulate, remaining alive but in a dormant and extremely tough low-energy state. We might conceptualise probiotic materials in this way where beneficial strains remain viable in indoor environments but are constantly responding and circulating through conditions of germination and sporulation, providing beneficial agency only when specific conditions arise. The role of design in this sense offers the potential to stimulate the bacteria to undergo this transition, where they can then remain, ready to be reanimated in response to a trigger – such as the presence of water due to flood or damp, or the presence of a pathogen.

We sought to build on these strategies. *B. subtilis* strains have the ability to survive in dry indoor conditions by transitioning to these dormant states. This sporulation process is well understood, and bacterial spores have been observed to lie dormant indefinitely. They are able to withstand extremes in temperature, pressure, pH and UV light. We were keen to see whether we could demonstrate this phenotypic transition from motile single cells to a surface-based biofilm, and finally sporulation within our material matrix. This bioaugmentation of the substrate/organism interface drives the development of a resilient probiotic material. We were keen to see whether a microbial-led material strategy could facilitate the self-assembly of *B. subtilis* biofilms within the material matrix to create a bioanimate material resilient enough to remain viable in normal indoor environments without the need for an additional supply of nutrients or water.

SEM analysis at specific timepoints allowed visual observation of this transition within the material matrix. This appeared to show evidence of both biofilm formation and a phenotypic change in state from germinated cells to spores over time. Figure 3.6 shows samples under SEM. At days 1 and 7 the *Bacillus* samples appeared still in their germinated state; by day 21 they appeared as spores which were successfully regerminated after 28 days. Based on the known longevity of spores, it can be reasonably expected that survival and subsequent regermination would be possible after indefinite storage under normal indoor conditions.

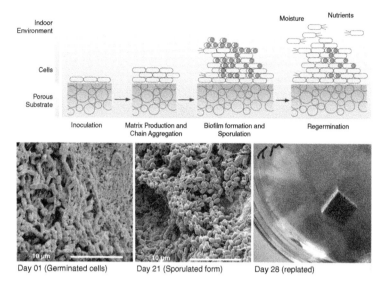

Indoor Environment

Moisture    Nutrients

Cells

Porous Substrate

Inoculation | Matrix Production and Chain Aggregation | Biofilm formation and Sporulation | Regermination

Day 01 (Germinated cells)    Day 21 (Sporulated form)    Day 28 (replated)

Figure 3.6 SEM observation of the sporulation process in the material matrix

To explore this long-term survival further, a quantifiable methodology based on cell counting was developed using a comparison of cells in to cells out. During the inoculation stage, a known number of cells were introduced into the material samples and a cell-counting method was developed to compare the number of cells extracted from the material samples. To do this, we needed a way to extract the cells from the solid material matrix. Cube samples were placed in a bespoke crushing instrument made for the test to break apart the materials, providing access for the microbes. Crushed samples could then be washed out and vortexed to separate the biological cells from the material particles. These cells were then replated and grown on nutrient plates. Using serial dilutions, cell counting was then used to determine the number of viable cells, which we could compare to the number that were put in (Figure 3.7). Using biological triplicate studies under sterile conditions, samples were tested at specific timepoints over the course of a month.

Figure 3.7 Process of quantifying microbial survival in probiotic materials

Micro-Scale: Probiotic Materiality

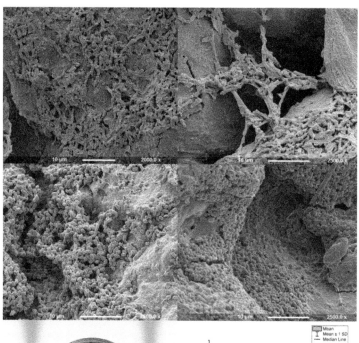

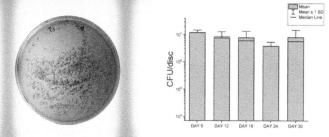

Figure 3.8 Survival of *B. subtilis* in ceramic material over one month

Results showed no significant difference in the number of cells counted at each timepoint. From this we can conclude that *Bacillus subtilis* cells inoculated into the ceramic material were able to survive for one month under normal indoor conditions without nutrient or water restock (Figure 3.8).

### Stage 3: Probiotic Action
The final stage of work at the micro-scale set out to assess and validate the probiotic action of the hybrid material. This was defined

specifically through the ability of the living material to inhibit the growth of a known pathogen. The beneficial relation to health here follows the hypothesis that probiotic materials could reduce the pathogen load in a building, thereby reducing the likelihood of surface-acquired infection.

In an initial test, a disc of the living material was placed on a plate cultured with MRSA. An inhibition ring can be observed around the disc of material in Figure 3.9, which offered evidence in support of the hypothesis that the probiotic material could inhibit the growth of an AMR pathogen. In an attempt to quantify this, triplicate inhibition assays were used to determine the ability of the probiotic material to inhibit MRSA. Three combinations were tested, including co-culture: *B. subtilis* + *S. aureus*; mono-cultures, *B. subtilis* + tryptic soy broth

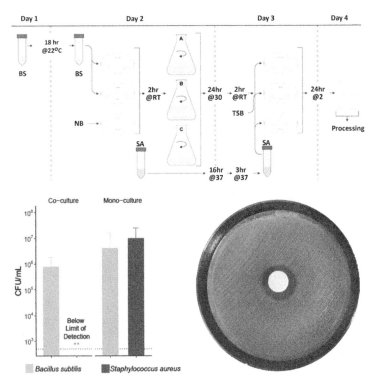

Figure 3.9 Top – methodology for quantifying pathogen inhibition of *Staphylococcus aureus* by probiotic materials; bottom left – graph showing co-culture inhibition; bottom right – inhibition ring

(TSB); and novobiocin (NB) + *S. aureus*. The data showed that *S. aureus* cells were below the limit of detection on the co-culture plate (with controls), suggesting that the probiotic material was able to inhibit the growth of MRSA (Figure 3.9, top and bottom left).

The work shows that benign bacteria can be integrated into architectural materials (probiotic materials), survive over time and prevent AMR bacteria colonisation.

## CONCLUSION: A PROBIOTIC PARADIGM

The work here serves as a fascinating proof of a concept for creating probiotic materials for buildings. It shows that building materials embedded with beneficial, healthy, living functions can survive for extended times without maintenance or the need for water or nutrient restock. However this must also be acknowledged as a relatively small piece of work that raises a multitude of further questions before such materials might be used in buildings. I refer to the technology readiness scale that was referenced in the first book in this series, *Living Construction* (Dade-Robertson, 2020). Although this research sits at the lower end of this scale, it also means that the many questions raised provide fascinating next steps in this area that hopefully this work can develop and inspire.

Some of these questions raised inform the work in the next two chapters, while others will need to be addressed in work that follows. At the time of writing we are just embarking on a follow-on project from this research. In collaboration with Jack Gilbert's lab at the University of California, San Diego, this project will explore in further depth how specific material properties at the micro-scale of design can influence mechanisms of gene expression and metabolites that regulate germination and the ability to out-compete pathogens. It will also explore genetic manipulation strategies for novel strain engineering of *B. subtilis* to augment the action and resilience of these mechanisms. From an architectural perspective, we will explore how these living materials might be designed as architectural scale objects, surfaces or components, and question how they could be integrated into built environments as interventions in order to test these innovations. These will be developed not only in line with physical concepts, including architectural space and material robustness and fabrication, but also in line with new guidelines for ethical and equitable implementation of these new approaches in buildings.

Design of interventions in this way requires a shift in scale of the design focus. Here the work must operate in consideration of the material upwards and question how microbial interventions can be informed by indoor environmental conditions. New methodologies will be needed to take these approaches out of the laboratory setting and explore their impact on the spaces and buildings in which they will operate. Key challenges will be to explore whether the microbes within the material will be able to have any agency outside of the matrix. If we can show that they can be shed, or transferred in some way, it raises the possibility that probiotic material interventions could serve as direct sources of beneficial microbes for other materials, surfaces and spaces within buildings. In this way, probiotic materials can be used to develop interventions that can inform the microbiome of buildings. In the same manner, they may also be able to impact the human microbiome and potentially could contribute to holobiont health. It is to these challenges and concepts we turn in the next chapter.

**REFERENCES**

Blaser, M. J. (2016). Antibiotic use and its consequences for the normal microbiome. *Science, 352*(6285), 544–545.

Cruz, M., & Beckett, R. (2016). Bioreceptive design: a novel approach to biodigital materiality. *Architectural Research Quarterly, 20*(1), 51–64. https://doi.org/10.1017/S1359135516000130.

Dade-Robertson, M. (2020). *Living construction*. Routledge.

Duc, L. H., Hong, H. A., Barbosa, T. M., Henriques, A. O., & Cutting, S. M. (2004). Characterization of *Bacillus* probiotics available for human use. *Applied and Environmental Microbiology, 70*(4), 2161–2171.

Fahimipour, A. K., ben Mamaar, S., McFarland, A. G., Blaustein, R. A., Chen, J., Glawe, A. J., Kline, J., Green, J. L., Halden, R. U., van den Wymelenberg, K., Huttenhower, C., & Hartmann, E. M. (2018). Antimicrobial chemicals associate with microbial function and antibiotic resistance indoors. *mSystems, 3*(6). https://doi.org/10.1128/msystems.00200-18.

Gibbons, S. M. (2016). The built environment is a microbial wasteland. *mSystems, 1*(2), e00033-16. https://doi.org/10.1128/mSystems.00033-16.

Gibbons, S. M., Schwartz, T., Fouquier, J., Mitchell, M., Sangwan, N., Gilbert, J. A., & Kelley, S. T. (2015). Ecological succession and viability of human-associated microbiota on restroom surfaces. *Applied and Environmental Microbiology, 81*(2), 765–773.

Hegarty, B., Dannemiller, K. C., & Peccia, J. (2018). Gene expression of indoor fungal communities under damp building conditions: implications for human health. *Indoor Air, 28*(4), 548–558.

Hu, J., ben Maamar, S., Glawe, A. J., Gottel, N., Gilbert, J. A., & Hartmann, E. M. (2019). Impacts of indoor surface finishes on bacterial viability. *Indoor Air, 29*(4), 551–562.

Kampf, G. (2018). Biocidal agents used for disinfection can enhance antibiotic resistance in gram-negative species. *Antibiotics, 7*(4), 110. https://doi.org/10.3390/antibiotics7040110.

Karana, E., Blauwhoff, D., Hultink, E., & Camere, S. (2018). When the material grows: a case study on designing (with) mycelium-based materials. *International Journal of Design, 12*(2), 119–136. http://www.ijdesign.org/index.php/IJDesign/article/view/2918.

Kennedy, T. A., Naeem, S., Howe, K. M., Knops, J. M. H., Tilman, D., & Reich, P. (2002). Biodiversity as a barrier to ecological invasion. *Nature, 417*(6889), 636–638.

Lax, S., Cardona, C., Zhao, D., Winton, V. J., Goodney, G., Gao, P., Gottel, N., Hartmann, E. M., Henry, C., & Thomas, P. M. (2019). Microbial and metabolic succession on common building materials under high humidity conditions. *Nature Communications, 10*(1), 1–12.

Li, S., Yang, Z., Hu, D., Cao, L., & He, Q. (2021). Understanding building–occupant–microbiome interactions toward healthy built environments: a review. *Frontiers of Environmental Science & Engineering, 15*(4), 1–18.

Lorimer, J. (2020). *The probiotic planet: using life to manage life.* University of Minnesota Press.

Mahnert, A., Moissl-Eichinger, C., Zojer, M., Bogumil, D., Mizrahi, I., Rattei, T., Martinez, J. L., & Berg, G. (2019). Man-made microbial resistances in built environments. *Nature Communications, 10*(1), 1–12.

Perini, K., & Rosasco, P. (2013). Cost–benefit analysis for green façades and living wall systems. *Building and Environment, 70*, 110–121.

Rook, G. A. W., & Bloomfield, S. F. (2021). Microbial exposures that establish immunoregulation are compatible with targeted hygiene. *Journal of Allergy and Clinical Immunology, 148*(1), 33–39.

Setlow, P. (2014). Germination of spores of *Bacillus* species: what we know and do not know. *Journal of Bacteriology, 196*(7), 1297–1305.

Vasudevan, R., Kennedy, A. J., Merritt, M., Crocker, F. H., & Baney, R. H. (2014). Microscale patterned surfaces reduce bacterial fouling-microscopic and theoretical analysis. *Colloids and Surfaces B: Biointerfaces, 117*, 225–232.

# Meso-Scale: Probiotic Design Interventions

## Chapter 4

DOI: 10.4324/9781003207917-4

In the previous chapter I outlined an approach to develop probiotic materials through design strategies and methodologies operating at the micro-scale of the microbe–material interface. Through the focus on antimicrobial resistance (AMR), these materials serve as proof of the concept that benign microbes can be integrated in building materials, survive and remain viable in indoor environments over time without maintenance. They can serve to inform healthier built environments by limiting the presence of pathogens; and they may also make it possible for this approach to integrate and reintroduce other benign microbes into buildings that are beneficial for holobiont health through their potential to shape the human microbiome.

In this chapter I want to address an approach for probiotic design to align with the second example of the contemporary pathologies offered in Chapter 2, that of diseases of 'missing microbes'. These diseases describe the increase in chronic and immune-related pathologies observed over the last 50 years that have been associated with a lack of microbes in the contemporary urban environment, especially inside buildings. These findings spark the idea for novel probiotic interventions that can directly reintroduce these missing communities of microbes into buildings through the material condition. Diseases of missing microbes highlight not only the need for microbial diversity in buildings, but also the need for these microbial communities to impact the human microbiome of the buildings' occupants. To ensure this exposure, probiotic interventions will depend on the agency of these microbial communities acting outside of the material condition. To approach this, we can build on existing knowledge about microbial transmission of pathogens to humans in buildings, such as via touch and inhalation.

Exploration of this approach will require conceptual and methodological shifts as well as an upwards shift in scale of the design focus. Whereas the strategy in Chapter 3 explored the role of a single strain of good bacteria to inhibit the presence of bad bacteria, this chapter will focus more on microbial communities containing multiple strains of bacteria, other microbes and their metabolites. Conceptually, the material condition should be imagined not only as a niche for beneficial microbes but also as a source of beneficial microbes for other parts of the building. Design will play a key role here in augmenting notions of microbial shedding and transference of microbes out of the material matrix. Regarding the required shift in scale, in this chapter I move upwards from small material cubes to the scale of architectural objects, surfaces and components in order to explore how probiotic materials might be designed as building elements. This raises methodological challenges as to how to fabricate objects at larger scales that can still operate within the confines of laboratory equipment and the microbiological methodologies.

The design focus here also shifts from the micro-scale of the 'microbe–material interface' to the meso-scale operating at the 'material–built environment interface', and questions how these factors relate to the microbiome of a building space. At this scale, objects can be designed as interventions that can be installed in buildings and then tested in terms of their impact on the microbial composition of the surrounding walls, floors, surfaces and objects. Finally, design and testing in this way requires a new methodological approach. This approach moves from a laboratory-based culturing methodology, working with and testing against single strains, to one that engages with wider communities of microbes and their agencies within a real-world building environment. The work described in this chapter attempts to engage a probiotic design agenda with the microbiome of the built environment, and develops a methodology for engaging DNA sequencing approaches as a measure of design intent.

## MISSING MICROBES

Chapter 3 looked at the beneficial role that microbial diversity can play in buildings by offering mechanisms of microbial competition. This diversity was positioned as a kind of shield that prevents the unchallenged growth of particular strains on surfaces. Another beneficial role that microbial diversity can play in health relates to holobiont health and formation of the human microbiome.

Diseases of missing microbes (Blaser, 2014, p. 288) describe pathologies resulting from the absence of particular microbes from our bodies and daily lives that is causing varying states of dysbiosis throughout urban populations. Specifically, these 'epidemics of absence' (Velasquez-Manoff, 2012) describe the observed rise in chronic, immune-related illnesses – including asthma, diabetes and multiple sclerosis, among others – in the early 21st century. While infectious diseases have dropped, cases of immune-related diseases have risen drastically since WWII, as shown in Figure 4.1.

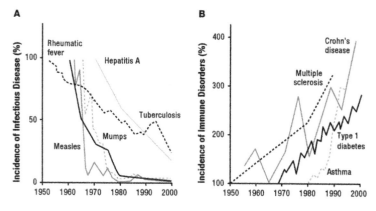

Figure 4.1 Disease trends, 1950–2000 (Bach, 2002)

These new pathologies are more predominant in urban than rural areas as urbanization reduces the transfer of diverse environmental microbiota indoors (Parajuli et al., 2018). Urban dwellers face reduced exposure to environmental microbiota, which negatively impacts normal immune function (von Hertzen & Haahtela, 2006). This involves socio-ecological aspects based on the sciences of immunoregulatory diseases, where the focus is on a lack of relevant microbial exposure that is necessary to regulate immune function. As well as the physical diseases described, missing gut microbes are also being related to mental health conditions such as schizophrenia, bipolar disease and autism.

In Chapter 2 we used these epidemics of absence to challenge the ontological narrative of microbes solely as pathogenic, disease-causing agents. Instead, we saw how they can be used to reframe microbes as a fundamental part of the human body, a co-evolved, symbiotic entanglement of cells and genes – old friends that we have failed to keep in touch with. We summarised this to challenge the idea that to design healthy spaces we should eradicate microbes. But, perhaps even more provocatively, by their very nature these diseases suggest that healthy spaces should contain microbes – rich, diverse communities of them. In urban buildings, especially where the surrounding environmental microbiodiversity is low, we may need to explore novel design interventions that purposely seek to add or restore these microbes and their necessary niches in buildings. If we can understand how these interventions can impact the microbiome, we can then explore how design can begin to inform a healthier, more diverse indoor microbiome.

## INFORMING A HEALTHY INDOOR MICROBIOME

The question of exactly what a healthy indoor microbiome is or might comprise, especially in terms of what types of microbes are desirable or what amount of diversity is necessary for impact, is still under investigation. What is clear at this point, however, is that, in urban environments, contemporary buildings can be understood as problematic through their lack of microbial diversity. As a result of antibiotic mentalities operating at multiple scales across the city, indoor microbiomes lack the types or sufficient diversity of microbes that are necessary for healthy immune development and maintenance of the human microbiome. While we await further understanding from the medical fields as to the specifics of these human/microbial health relationships, we can start to explore the potential for design strategies and interventions to begin to inform or shape the indoor microbiome in certain ways. Figure 4.2 envisages probiotic materials acting as a beneficial layer of microbes for the indoor environment.

To begin, we might start to target what a healthy indoor microbiome might be, in line with the 'Old Friends' hypothesis

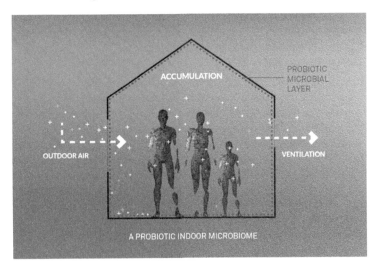

Figure 4.2  A probiotic indoor microbiome

(Rook et al., 2013). This suggests how, from an evolutionary perspective, the microbial inputs necessary for health should be understood as those associated with the natural environments alongside which humans have evolved. Primarily these are environmental microbes associated with nature. In relation to buildings, Rook et al. expand on this through the evolution of human dwellings, their materiality and the resultant microbiomes they inform. The authors discuss how pre-modern dwellings, constructed using natural materials, would have shaped subsequent microbiota that would not have differed much from those of nature. Structural elements made of mud, stone or timber were then covered with thatch or turf. Material finishes would have contained soil, clay and animal dung. Rook et al. contrasts these to modern homes built mostly from synthetic products and with tightly sealed envelopes. Modern buildings can thus be seen to inform microbiota that bear no resemblance to nature, especially in urban environments dissociated from natural areas. These 'unnatural' microbiota are seen as not sufficient or optimal for structuring the acquired parts of the human microbiome in early years, especially those necessary for normal immune development (Rook & Bloomfield, 2021). A comparable differentiation is observed between urban and rural buildings. Dwellings in rural areas tend to exhibit more diverse microbiomes than dwellings in urban areas as they allow microbiota from the natural environment around them to move indoors. Broadly then, we can summarise thus: dwellings that contain microbiomes more similar to those in nature are likely to be beneficial for the holobiont.

A second line of enquiry that we might use to inform what a beneficial indoor microbiome might look like builds on the so-called 'farm effect'. This describes the literature observation that communities living close to farms appear to exhibit fewer instances of immune disorders such as asthma and other allergic diseases (Ludka-Gaulke et al., 2018). This has been largely explained as a result of early life contact with farm animals and their microbes and resulting immune mediation (Riedler et al., 2001). These cases have been highlighted through microbiome research undertaken with farming communities in the US. The interesting nuance in this evolving story relates to stark differences in asthma prevalence between two similar farming populations of Amish and Hutterite children. Amish children are observed to have a significantly lower prevalence of asthma and allergic sensitisations compared to

Hutterite children – despite the fact that these two communities share many similarities in their cultural and genetic histories and lifestyles. The relevant difference appears to be related to the different farming methods used and their effects on shaping the microbiomes of the adjacent home environments. Hutterite communities live on large farms and use industrialised technologies, while the Amish live on small dairy farms and practise traditional farming methods using horses. This was observed to have a significant impact on the microbial composition in house dust within the two communities. This offers insights into the mechanistic pathways for this phenomenon that suggest that microbial exposure in home environments that are rich in microbes associated with farms protects against allergic asthma by engaging and shaping innate immune pathways (Ober et al., 2017).

## INDIRECT INTERVENTIONS TO INFORM MICROBIODIVERSITY

The challenge for design interventions appears to be to try to shape indoor microbiomes in urban environments towards the kinds of microbiota associated with more natural environments, and possibly farms. Within the literature, 'indirect' methods, such as having pets (particularly dogs), have been observed to increase indoor microbiodiversity through their mechanisms of tracking outdoor microbes indoors (Fujimura et al., 2010). Building strategies have mostly focussed on façade strategies, particularly those that permit opening windows to increase air exchange, and the use of HVAC and other ventilation systems that utilise fresh air – both of which are understood to increase the presence of outdoor microbes on surfaces and in the air. While in support of these approaches, certain limitations exist. These strategies must be balanced in line with contemporary energy issues to prevent unnecessary heat leakage. Also of relevance here is the quality of the outdoor air. In highly polluted areas, common in cities, it may be more detrimental to health to have windows open. In other areas, such as heavily urbanised parts of cities where there is little to no green space, the outdoor microbiodiversity may be so limited that there may be no beneficial result to the indoor microbiome.

Indirect strategies then are dependent on the quality of the environmental microbiodiversity around them. Increasing microbiodiversity in cites in line with greater biodiversity, especially in smaller areas around buildings, will be a fundamental part of creating probiotic built environments. In some ways, we know mostly how to do this. More green and natural spaces in cities, especially if targeted towards biodiversity, will provide sources of microbial diversity for surrounding buildings. However, in a similar manner to existing urban greening approaches, challenges to implementing this across cities are more political and economic than technical. These strategies will be discussed in the next chapter; but these limitations highlight the need to also explore what we will call 'direct' strategies to inform more diverse indoor microbiomes.

## DIRECT INTERVENTIONS TO INFORM MICROBIODIVERSITY

Direct interventions describe strategies that seek to purposely add microbes to buildings as a means of increasing the diversity of the indoor microbiome. This approach might be comparable to landscape rewilding strategies that look to reintroduce keystone species into dysbiotic ecologies to restore healthy function. Here the building is considered as the dysbiotic ecology, and so the practice of introducing and distributing species begins to engage with spatial strategies. Referring again to the probiotic practices identified by Lorimer (2020), strategies of distribution relate to the notion of reclaiming and securing territories and restoring vectors of movement that have been lost due to previous antibiotic mentalities. While at the scale of landscapes these terms relate to large areas of habitat and connectivity between them, the use of the same terms for microbial rewilding of buildings might infer smaller scales of operation. Structuring of microbial territories and niches in buildings can be achieved by implementing materials that are able to support colonisation. Bioreceptive building surfaces, components and objects can serve as significant territories for microbial presence in indoor environments. In a passive manner, the porosity of these materials can facilitate the collection of outdoor microbes along with other organic compounds present from activities in buildings (Lax et al., 2019). In addition, they can retain moisture at the surface interface (Hoang et al., 2010). If these materials are made from or directly inoculated with environmental microbial communities, they potentially become sources of diversity for the indoor microbiome around them.

The practice of designing microbial materials, specifically from the perspective of a biodesign agenda, has a key role to play here. Work on materials that are manufactured using bacterial strains have been explored for use in buildings (Dosier, 2014), though in this case the organisms were not alive at the time of implementation. Other work has explored integrating bacterial cells in concrete materials that enable self-healing functions throughout the building's lifetime (Jonkers & Schlangen, 2010). Although these strains remain 'alive' in a sporulated state, little is known about their impact on the microbiome or their agency outside of the material.

In Chapter 2 we read how material surfaces in buildings have been characterised as 'microbial wastelands' from indoor microbiome research (Gibbons, 2016). However, this current state of knowledge has been informed by studies on common materials. These mostly include materials that have been developed over the last 70 years in line with the antibiotic mentalities described in Chapter 1 which have sought to minimise microbial colonisation. The wealth of experimental or more fringe materials that are central to the expanding fields of bio-integrated design and living architecture, by their very nature, would be expected to inform significantly different indoor microbiomes and surface conditions from those currently characterised. For example, mycelium materials, bacterial celluloses, algae-laden hydrogels and DIY biopolymers have become common material currency across the growing number of researchers and academic courses in this field (Camere & Karana, 2018; Malik et al., 2020; Zolotovsky, 2017), yet their potential to inform the microbiomes of the buildings they are intended for remains mostly unknown and untested.

The ability of bioreceptive and biomaterials to support living species could significantly contribute to microbial diversity in buildings. In the student project Loofahtecture, fully grown, biohybrid building components were developed using a biolamination process combining loofah and mycelium materials.[1] These were then explored as scaffolds to host the growth of multiple living species. Shown in Figure 4.3, on the front surface the micro-fibrous nature of the loofah provided a bioreceptive niche for cyanobacteria and soil microbes (pioneer species) as a precursor to the succession of larger organisms. The back surface of the mycelium provides thermal performance while still allowing for the growth of fungal hyphae under selective conditions. Building materials constructed and assembled using grown or living materials could provide valuable sources of microbiodiversity for the built environment.

The application and commercial development of these experimental materials have typically been limited by preconceptions

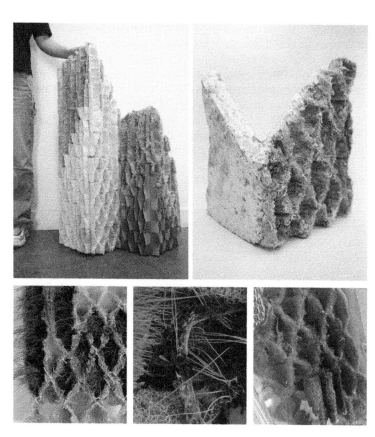

Figure 4.3 Loofah/mycelium hybrid architectural components

regarding structural performativity, material longevity and their unconventional or taboo aesthetics. Materials that exhibit moisture, that are soft or sticky or that decay (or grow) are mostly rejected for real-world application. Yet these properties are fundamental to the many of the natural materials alongside which the holobiont has evolved (and those used to build in pre-modern times). The potential for these materials and systems to provide ecosystem health for cities and humans through the microbiomes they inform can add other lines of enquiry for these materials and those convinced of the need for radical change in the way we build and construct future cities. For biodesigners, methodological challenges and access to the knowledge and technology of metagenomics are likely to be barriers to research in this area. Collaboration with environmental microbiologists at this stage of knowledge is necessary; and, despite reductions in sequencing costs, engaging with microbiome studies is still likely to require some form of funding. Collaboration with other researchers in the fields of building science, as well as with clinicians and immunologists, would be valuable to begin to understand the role these interventions could play in shaping human microbiomes in the built environment. It should be stated that there is no ideal or agreed assessment for these questions yet.

## FROM 'MICROBIAL TRANSMISSION' TO 'MICROBIAL ENTANGLEMENTS'

A fundamental question in this area is to explore at this stage whether probiotic design interventions could impact the microbiome of a building, and potentially the human microbiomes of its occupants. Based on understanding of how pathogens are transmitted in buildings, it is likely they could. Conceptually, probiotic materials in this way might be imagined as beneficial 'fomites'. The term fomite is typically used in a negative sense to describe objects or materials that, when contaminated with a pathogen, can then transmit disease to a human (Stephens et al., 2019). If materials contain communities of beneficial microbes, it would make sense that these can be transferred to a building's occupants through the very same mechanisms of touch, inhalation and ingestion. This requires a reframing of microbial transmission to what I will call here 'microbial entanglement', where the interventions and the way they are used are designed so as to facilitate these entanglements.

These entanglements should be understood as a designable condition that can inform new relationships between architecture and the holobiont body. Whereas previous imaginaries of the body have informed architectural notions of beauty and proportion (in the case of the classical body) or standardisation (in the case of the modern body), these bodies (and their architectural explorations) were not aware at the time of their microbial symbionts. The holobiont body might inform design strategies that promote a 'thickening' of the invisible microbial layers enveloping the indoor environment and occupant body. Exchanges between the architecture and the body in this sense become more than a physical bodily phenomenon such as tactility or physical/ergonomic relationships. Instead, this engagement is between architecture, microbes, the immune system and holobiont genes. These entanglements would also have economic and political considerations that manifest in the inner organs, nasal passages and dermal layers of the population. Design can play an important role in facilitating these entanglements, not only in raising many functional, operational considerations but also in ensuring that probiotic cities enable and facilitate equal access to beneficial microbes across the population, including among minority and lower socio-economic groups.

## PROBIOTIC INTERVENTIONS

At the meso-scale of operation, these factors can be considered in the design of probiotic surfaces, objects and components of buildings that can shape the indoor condition. In the later stages of the NOTBAD project we speculated with a series of probiotic wall tiles that could serve as a source of beneficial microbes for human entanglements. Their design sought to exploit microbial transfer via the mechanism of human touch by creating textural surface geometries that were both environmentally performative and aesthetically intriguing to encourage touch (Figure 4.4). These textural geometries sought to increase both touch frequency and duration. However they were also observed to encourage a rubbing movement as people's hands explored the tactility, a strategy that has been observed to increase microbial transfer from material to finger pads in contact studies on pathogen transfer (Julian et al., 2010; Zhao et al., 2019).

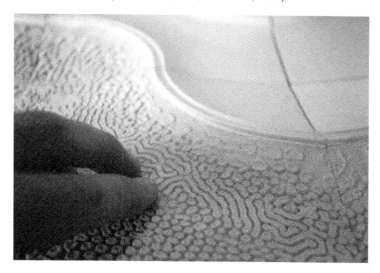

Figure 4.4  Probiotic tile textures

The second important role that geometry can play in facilitating microbial entanglement lies in forms and geometries that can facilitate the trapping or shedding of microbes from the material in relation to indoor airflow. In contrast to flat, smooth surfaces, three-dimensional and geometrical forms can be shaped to catch or trap outdoor microbes within the porous matrix. If the same mechanisms could facilitate shedding of microbes, probiotic interventions would have the potential to serve as sources of microbes to other surfaces or other parts of buildings. The role of airflow is well understood as a key mechanism in relation to the spread of viral particles in buildings, as well its effect on pathogen abundance in spaces (Kembel et al., 2012). Computational fluid dynamic (CFD) approaches are increasingly being used to investigate the prevention and control of pathogen transmission (Peng et al., 2020). In a similar manner they can potentially be used as design tools to promote beneficial exposure. Computational approaches that can incorporate airflow simulations can sculpt and optimise surface geometries according to criteria specific to a building environment (Figure 4.5).

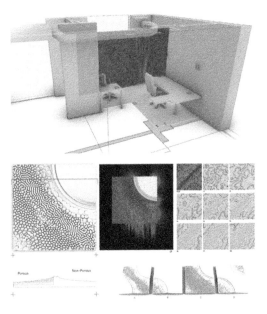

Figure 4.5 Computational fluid dynamic simulations to promote microbial capture, shedding and spatial translocation

The tile typologies were modelled computationally with geometrical differentiations to create microclimate variations on the surface in relation to airflow simulations. The peaks and valleys create areas of micro-turbulence and eddies as air flows past the surface seeking to resuspend and shed microbes to the air and surrounding surfaces.

The tiles also facilitate a spatial agenda to the planning and layout of probiotic materials as wall surfaces. This has both vertical and horizontal considerations that can relate to airflows, adjacency to doors, windows and other architectural features, as well as to bodily dimensions. Different microbial inoculants could be used in different areas. The microbiota in infant breathing zones is known to differ in composition from that in adult breathing zones (Hyytiäinen et al., 2018). Therefore, specific microbial inoculants could be targeted at crawling babies and toddlers, for example, to maximise entanglements, and others could be used for adults through mechanisms of touch, resuspension and respiration (Figure 4.6).

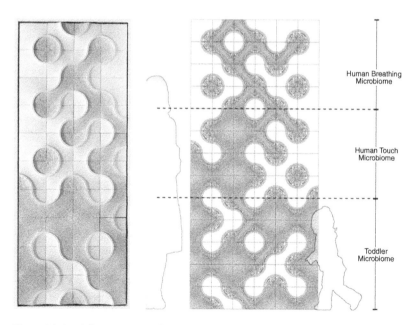

Figure 4.6 Spatial organization of probiotic agencies

In a second project CFD simulations of airflow of a test space were used as a design generator to inform the geometric form and niche conditions of the probiotic intervention. Airflow simulations can be imported into procedural design software, where, using a particle approach, areas of air/surface collisions can be determined. Surface scale extrusions and depressions and curves and folds can be informed in relation to the air flowing past and over them. These surface variations, and subsequently the geometries themselves, become locally and spatially programmable, designed to augment microbial deposition and resuspension with the aim of shedding microbes in response to airflow vectors and human touch. Airflow simulations incorporated a time-based process whereby each second of the simulation contains a dataset that then be used to inform a series of computational 'frames' containing geometrical boundaries (Figure 4.7).

Here a computational methodology was developed whereby these frames inform layers of points through which vector lines can be connected to create a continuous 'curve' associated with those layers. A differential line growth algorithm was then used to create geometric variation across the layers, growing the geometry outwards. These layers define the notion of distance from the wall, which led to the creation of 'components' that become much more three-dimensional in form compared to wall tiles, offering much more sculptural area and form that can engage with the indoor environment.

**PROBIOTIC FABRICATION**
An important concept for the design of interventions at this meso-scale of operation relates to fabrication. In scaling up from material samples to architectural-scale objects the method of fabrication becomes challenged by physical factors and limitations of the scales associated with microbiological methodologies and equipment. These experiments were designed computationally, allowing for forms and geometries that facilitate microbial entanglements, but also enabling accuracy and repeatability during fabrication to allow for engagement with the microbiological methodologies and experimental testing.

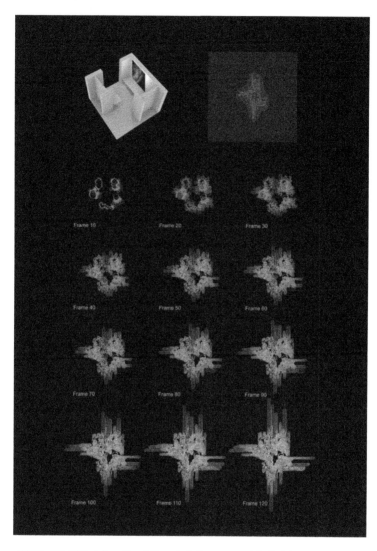

Figure 4.7 CFD airflow simulations to inform geometrical niches

In the cases here, tiles were chosen as a typology that permits large areas of wall to be covered using relatively small units. The 150×150mm size of the individual tiles meant they were still manageable within laboratory conditions, moving easily between autoclaves, chambers and fume hoods. To make them in an accurate and repeatable manner, moulds were produced form 3D printed masters and then cast according to mix designs from the micro-scale methodologies. A multilayer approach was used to define areas of porosity that would contain microbes and other areas of non-porosity that offer structural strength but also assisted variations in pattern tiling. When it came to producing multiple tiles for intervention testing, the workflow meant we could take batches of tiles to the lab for sterilisation and inoculation prior to experimental testing. A limitation less considered was the time taken to inoculate larger numbers of tiles as manually pipetting the microbial inoculant on to the tile zones was extremely time consuming and would not be feasible for producing multiple tiles to cover large wall areas. Spraying was discussed but brings risks associated with aerosolising microbes, especially if done on-site rather than in the laboratory, and so was discounted. To address this problem we developed an automated approach using a syringe pump and extruder attached to a robotic positioning system (Figure 4.8).

In this project we built on these lessons, and so the design of the components was informed by the need to develop an integrated bio-fabrication approach from the beginning, built around a robotic fabrication process. The development of an automated robotic

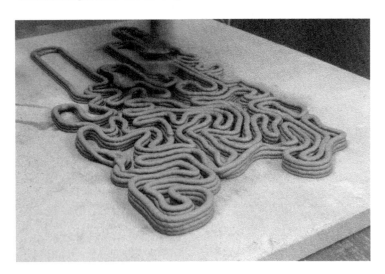

Figure 4.8 Robotic ceramic extrusion

workflow meant that the size of the individual components was no longer limited by the laboratory equipment. This not only meant the pieces could be larger, but also allowed forms and geometries that were no longer limited by casting techniques. The curves generated through CFD simulations of the space described could then be used directly as toolpaths for a robotic fabrication process using clay extrusion. These processes allow the fabrication of individual components with multiple variations in size and geometry in a manner that is not limited by the need to make moulds.

Following clay extrusion, components were bisque fired to achieve a ceramic condition with targeted porosity values, and then second fired with glazing applied to the rear faces to provide structural and aesthetic properties (Figure 4.9). To complete the biofabrication

Figure 4.9 Probiotic ceramic components

process, this computational methodology allows for the toolpath data to be used as a way of inoculating the components after fabrication with the microbial communities. Using scale factors to compensate for shrinkage after firing, a syringe pump was employed with a robotic end effector. The components could then be 'robotically inoculated', allowing for accurate and repeatable application of microbial cultures into the components for experimental testing.

## PROBIOTIC INTERVENTION TESTING

To finish this chapter, I want to outline an experimental approach undertaken to explore the potential of these meso-scale design interventions to impact the microbiome of a space. This defined the research question: 'Can a probiotic building prototype act as a viable biodiversity-enhancing intervention to provide a more diverse indoor microbiome?'

The hypothesis is that probiotic interventions can serve as a source of diverse environmental communities of microbes which could then be shed from the material substratum to provide beneficial agency to the microbiome of a building. If this could be demonstrated, together these mechanisms could be designed with the grander aim of designing healthy indoor microbiomes through increased diversity of an indoor microbiome, towards one comprising a higher constitution of outdoor, environmental microbes. We sought to answer this fundamental question through an experimental study. The starting point again became a question of which microbes we might design for.

## MICROBIAL COMMUNITY IDENTIFICATION

In the last chapter, the identification of *which* bacteria to integrate into materials was driven by the selection of individual strains exhibiting mechanisms that could inhibit pathogenic microbes. As we look now to 'communities' of microbes that are beneficial for human health, identifying specific strains of bacteria for this agenda proved to be a somewhat more challenging question. While it is well understood that not all bacteria are pathogenic, exactly what constitutes healthy bacteria for holobiont health is still not clear from a medical perspective. Research into 'good' bacteria is still relatively new in the field of microbiology as so much work over the last decades has focussed on understanding pathogens. Building on the literature, and in line with the Old Friends theory, the decision was made to explore communities of soil microbes. Soil would be collected from a rural location, expected to be relatively rich in its diversity of microbes, and then isolated and used for inoculation into the probiotic surface components. We could then configure the experiment to monitor the effect this installation in a test space by measuring the microbiome using 16S sequencing.

## MICROBIAL SURVEY

The test space for the study became my office in the School of Architecture. Using guidance from the literature we began with a microbial survey of the space through a mix of culturing and metagenomic approaches using 16S sequencing. In line with other types of survey common to architecture, this data would provide an understanding of the 'existing' microbial condition of the space at that time. This provides not only a characterisation of the microbiome but also a condition to test against and subsequently determine whether the probiotic intervention could impact its constitution. The study was planned using a before, during and after approach, which meant we could monitor how the microbiome of the space changes from its initial condition when the installation was introduced and how it behaves over time. Similarly, the effect of removing the intervention after a certain period could be monitored by continuing to take measurements to understand how the microbiome responds afterwards. Alongside the microbial samples we monitored the environmental conditions of the space (including temp/RH%) and documented how the space was occupied and used using during the course of the experiment. Finally, architectural approaches of drawing and modelling were used to helps us understand the built environment conditions of the space locally, including the HVAC system controlling these conditions in the space (Figure 4.10).

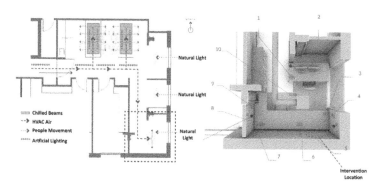

Figure 4.10 Engaging probiotic design with the microbiome

Beyond a pure characterisation of the microbes that exist in a space, from a design perspective the survey can help us understand the role the built environment is playing in the microbial constitution. These conditions can then be used as a design driver for interventions aimed at manipulating this constitution towards a healthier condition. On one hand, we might envisage whole walls covered in probiotic material. On the other hand, if we can show that the microbes in the material are able to shed from the material to other surfaces or parts of the space, smaller interventions or wall areas might suffice. In this early stage of the research, cost limitations, both in terms of manufacturing the probiotic interventions and the associated costs of sequencing, meant that the work would operate on the concept of a smaller intervention, and to assess if it could affect the microbiome of the test space.

**INTERVENTION LOCATION**

The question of where to install the intervention was informed by the CDF simulations discussed above, located to try to maximise its potential to influence the microbiome of the space. If we return to the question of whether the microbes within a probiotic intervention can be transferred to other parts of the test space, this approach sought to take advantage of this airflow in order to create turbulence at the material–environment interface, which can aid processes of microbial resuspension, shedding and spreading to other parts and surfaces in a building.

Based on these simulations, importance was given to a vertical wall area that was observed to receive direct airflow from an air-conditioning fan shown in Figure 4.10. Based on this location, a series of sampling points were then identified at specific points around the space to provide metadata through which 16S sequencing data could be positioned. These included a mix of horizontal and vertical surfaces (such as walls, desks and the floor), and comprised a range of materials (plaster/paint, plywood and concrete). These were then spatially located at varying distances form the installation to try to explore how far microbes were travelling around the space.

**MICROBIAL SAMPLING**

The study ran for nine weeks, seeking to achieve a notion of longitudinal data as advised in the literature, though within the confines of the funding for the work. We employed an approach that sampled the ten location sites once a week for three weeks to establish the existing conditions. Following installation of the intervention, the same sites were then sampled for three weeks, after which the intervention was removed and sampling continued for another three weeks. This informed a before, during and after approach through which we could structure our metagenomic results. Sampling locations were defined by $10cm^2$ markers to ensure repeatability, and samples were taken with cotton swabs using a repeatable methodology. They were then analysed using a 16S Illumina amplicon protocol designed to amplify prokaryotes (bacteria and archaea) using paired-end16S community sequencing (Caporaso et al., 2018).

## CONCLUSION

The results of this study will appear in a separate publication, where the specifics of the sequencing data will be presented and discussed. Briefly, however, the work serves as a first attempt to use a direct design intervention to inform the indoor microbiome, and hopefully can serve as a basis for further research in this area. It was clear that the intervention did serve to modify the microbiome of the space, with clear differentiation in the microbial composition of the space following installation of the intervention. It was also clear that the microbes in the ceramic components were shed and transported around the space, with microbes from the intervention found on other sampling points.

While these results need to be discussed within the context of the experiment, they serve again as proof that direct probiotic design interventions can inform the microbiome of a space. While it is a very small step, this work has the potential to develop further as we learn more about how these microbes behave over time and whether they can offer any beneficial mechanisms towards health, either by limiting pathogens or in relation to holobiont health.

## NOTE

1 Loofahtecture was a project undertaken in RC7 in 2021/22 with students Jinghui Wei, Junjie Lyu and Yuchen Qiu.

## REFERENCES

Bach, J.-F. (2002). The effect of infections on susceptibility to autoimmune and allergic diseases. *New England Journal of Medicine, 347*(12), 911–920.

Blaser, M. J. (2014). *Missing microbes : how killing bacteria creates modern plagues*. Oneworld.

Camere, S., & Karana, E. (2018). Fabricating materials from living organisms: an emerging design practice. *Journal of Cleaner Production, 186*, 570–584. https://doi.org/10.1016/j.jclepro.2018.03.081.

Caporaso, J. G., Ackermann, G., Apprill, A., Bauer, M., Berg-Lyons, D., Betley, J., Fierer, N., Fraser, L., Fuhrman, J. A., & Gilbert, J. A. et al. (2018). EMP 16S Illumina amplicon protocol V.1. Earth Microbiome Project. https://www.protocols.io/view/emp-16s-illumina-amplicon-protocol-kqdg3dzzl25z/v1

Dosier, G. K. (2014). Methods for making construction material using enzyme producing bacteria. Google Patents. https://patents.google.com/patent/US20110262640A1/en.

Fujimura, K. E., Johnson, C. C., Ownby, D. R., Cox, M. J., Brodie, E. L., Havstad, S. L., Zoratti, E. M., Woodcroft, K. J., Bobbitt, K. R., & Wegienka, G. (2010). Man's best friend? The effect of pet ownership on house dust microbial communities. *Journal of Allergy and Clinical Immunology, 126*(2), 410–412.

Gibbons, S. M. (2016). The built environment is a microbial wasteland. *mSystems, 1*(2), e00033-16. https://doi.org/10.1128/mSystems.00033-16.

Hoang, C. P., Kinney, K. A., Corsi, R. L., & Szaniszlo, P. J. (2010). Resistance of green building materials to fungal growth. *International Biodeterioration & Biodegradation, 64*(2), 104–113.

Hyytiäinen, H. K., Jayaprakash, B., Kirjavainen, P. V, Saari, S. E., Holopainen, R., Keskinen, J., Hämeri, K., Hyvärinen, A., Boor, B. E., & Täubel, M. (2018). Crawling-induced floor dust resuspension affects the microbiota of the infant breathing zone. *Microbiome, 6*(1), 1–12.

Jonkers, H., & Schlangen, E. (2010). Development of a bacteria-based self-healing concrete. In J. C. Walraven & D. Stoelhorst (eds), *Tailor made concrete structures* (pp. 109–109). CRC Press.

Julian, T. R., Leckie, J. O., & Boehm, A. B. (2010). Virus transfer between fingerpads and fomites. *Journal of Applied Microbiology, 109*(6), 1868–1874.

Kembel, S. W., Jones, E., Kline, J., Northcutt, D., Stenson, J., Womack, A. M., Bohannan, B. J. M., Brown, G. Z., & Green, J. L. (2012). Architectural design influences the diversity and structure of the built environment microbiome. *ISME Journal, 6*(8), 1469–1479.

Lax, S., Cardona, C., Zhao, D., Winton, V. J., Goodney, G., Gao, P., Gottel, N., Hartmann, E. M., Henry, C., & Thomas, P. M. (2019). Microbial and metabolic succession on common building materials under high humidity conditions. *Nature Communications, 10*(1), 1–12.

Lorimer, J. (2020). *The probiotic planet: using life to manage life.* University of Minnesota Press.

Ludka-Gaulke, T., Ghera, P., Waring, S. C., Keifer, M., Seroogy, C., Gern, J. E., & Kirkhorn, S. (2018). Farm exposure in early childhood is associated with a lower risk of severe respiratory illnesses. *Journal* of Allergy and Clinical Immunology, 141(1), 454–456. https://doi.org/10.1016/j.jaci.2017.07.032.

Malik, S., Hagopian, J., Mohite, S., Lintong, C., Stoffels, L., Giannakopoulos, S., Beckett, R., Leung, C., Ruiz, J., & Cruz, M. (2020). Robotic extrusion of algae-laden hydrogels for large-scale applications. Global Challenges, 4(1), 1900064.

Ober, C., Sperling, A. I., von Mutius, E., & Vercelli, D. (2017). Immune development and environment: lessons from Amish and Hutterite children. Current Opinion in Immunology, 48, 51–60.

Parajuli, A., Grönroos, M., Siter, N., Puhakka, R., Vari, H. K., Roslund, M. I., Jumpponen, A., Nurminen, N., Laitinen, O. H., & Hyöty, H. (2018). Urbanization reduces transfer of diverse environmental microbiota indoors. Frontiers in Microbiology, 9, 84.

Peng, S., Chen, Q., & Liu, E. (2020). The role of computational fluid dynamics tools on investigation of pathogen transmission: prevention and control. Science of the Total Environment, 746, 142090.

Riedler, J., Braun-Fahrländer, C., Eder, W., Schreuer, M., Waser, M., Maisch, S., Carr, D., Schierl, R., Nowak, D., & von Mutius, E. (2001). Exposure to farming in early life and development of asthma and allergy: a cross-sectional survey. The Lancet, 358(9288), 1129–1133. https://doi.org/10.1016/S0140-6736(01)06252-3.

Rook, G. A. W., & Bloomfield, S. F. (2021). Microbial exposures that establish immunoregulation are compatible with targeted hygiene. Journal of Allergy and Clinical Immunology, 148(1), 33–39.

Rook, G. A. W., Lowry, C. A., & Raison, C. L. (2013). Microbial "Old Friends", immunoregulation and stress resilience. Evolution, Medicine and Public Health, 2013(1), 46–64. https://doi.org/10.1093/emph/eot004.

Stephens, B., Azimi, P., Thoemmes, M. S., Heidarinejad, M., Allen, J. G., & Gilbert, J. A. (2019). Microbial exchange via fomites and implications for human health. Current Pollution Reports, 5(4), 198–213.

Velasquez-Manoff, M. (2012). An epidemic of absence: A new way of understanding allergies and autoimmune diseases. Simon & Schuster.

von Hertzen, L., & Haahtela, T. (2006). Disconnection of man and the soil: reason for the asthma and atopy epidemic? Journal of Allergy and Clinical Immunology, 117(2), 334–344.

Zhao, P., Chan, P.-T., Gao, Y., Lai, H.-W., Zhang, T., & Li, Y. (2019). Physical factors that affect microbial transfer during surface touch. Building and Environment, 158, 28–38.

Zolotovsky, K. (2017). Guided growth: design and computation of biologically active materials. MIT Press.

# Macro-Scale: Probiotic Cities

## Chapter 5

DOI: 10.4324/9781003207917-5

Moving upward again in scale, in this chapter I want to address probiotic design strategies for increasing microbiodiversity in buildings operating at the macro-scale of the building fabric to that of the city. As we look towards an imaginary of probiotic cities, these approaches and strategies must be applicable beyond the design of one-off buildings or façades. Instead, they will require strategies and design approaches not only to deal with a range of urban conditions but also in ways that seek to establish a network of microbial niches that connects the scale of the city all the way through to the microbes of the human gut. This incorporates soft niches in urban green spaces, hard niches of infrastructure and building envelopes right through to the meso-scale of indoor surfaces and the micro-scale of material porosity discussed in previous chapters. These scales must be understood as fundamentally interrelated, and probiotic cities would require strategies operating at all scales in order for microbiodiversity, including the holobiont body, to flourish.

At this city scale, probiotic design mentalities can align with existing urban greening, biophilic design, bioreceptive design and broader biodesign approaches, but with the distinction that such design strategies are prioritised through the notion of holobiont health and multispecies diversity as the fundamental target conditions. At times, this may be at odds with notions of aesthetics and maintenance of urban green spaces. They may not always align with species that are currently favoured for their ease of maintenance or their photosynthetic activity. The preferred scenario, however, is that strategies for holobiont health and climatic discourse can be considered together and perform in collaboration through the lens of creating healthy and resilient cities.

## REWILDING URBAN GREEN SPACE

Beginning at the scale of the city, urban environmental biodiversity can serve as a source of outdoor microbes for indoor environments. These include natural areas at the scale of large parks, down to gardens and small parcels of green areas around buildings. This involves not only green spaces but also brown and blue space in cities, and potentially even grey spaces that might be reconsidered potential urban surface areas to deliver beneficial microbial agency. As we read in Chapter 2, the indoor microbiome is in part populated by microbes entering from outside sources (Meadow et al., 2014), and so increasing the biodiversity of spaces around buildings can serve as an indirect method of increasing the amount and diversity of outdoor environmental microbes in buildings.

The idea that urban biodiversity can serve as an ecosystem service that is beneficial to human health in line with the holobiont concept has been proposed by various authors. These have been mostly understood and positioned through the role of landscape design and strategies that seek to radically increase the microbiodiversity of soft landscapes using a rewilding approach. The microbiome rewilding hypothesis of Mills et al. (2020) proposes rewilding existing urban green spaces from what are positioned as 'industrial urban' spaces to 'rewilded urban' spaces. These are characterised by a shift from current urban green approaches that inform low biodiversity urban habitats which are not immune protective towards high biodiversity urban habitats that provide immune-protective microbial exposure. Urban rewilding strategies are centred on the concept of the plant holobiont, which, if restored towards diverse communities, will stimulate beneficial cascades towards eventual diversification of the environmental microbiota. These mechanisms will also support food webs for encouraging greater diversity, including insects (especially pollinators) and larger birds and other animals.

In a similar manner, Microbiome-Inspired Green Infrastructure (MIGI), proposed by Robinson et al. (2021), describes a nature-centric urban infrastructure to enhance healthy human–microbe interactions. Building on a landscape-centred approach, MIGI promotes strategies of vegetation complexity, including the use of various native plants and specific soil types that are known to promote functional diversity of the environmental microbiome. Importantly, MIGI approaches are positioned in a way that considers the role of human interactions with these strategies in order to facilitate the microbial exposure necessary for immunoregulatory benefits. These include activities such as foraging and the need for hands-on engagement with MIGI features through maintenance and tending of vegetation. Robinson et al. further relate these landscape approaches to how they can inform building planning. This includes suggestions for how urban planning can take advantage of wind direction in relation to the airborne movement of microbes, and highlights the need for horizontal and vertical strategies to create continuous habitat connectivity across the urban fabric.

## MICROBIOTA TRANSLOCATION

In urban conditions where soil biodiversity is very low, it may be challenging to restore diversity through planting strategies. However emerging studies are showing evidence that revegetation can improve urban soil microbiota diversity by enhancing wild habitat conditions (Mills et al., 2020). A more direct approach in challenging cases would be to translocate microbial diversity from natural areas to urban areas. Translocation is already a common technique in landscape architecture for moving trees or soil, and in conservation and rewilding projects involving keystone species (Seddon, 2010). A similar approach can be explored specifically for microbiome health through translocating communities of soil microbes. This could involve translocating microbes themselves as a form of inoculant (Hui et al., 2019) or moving their environmental niche. Translocation of soil or forest materials into the built environment, for example, would also translocate the microbiome of that soil. This approach has been proposed by researchers in Finland exploring the effect of exposure to microbiodiversity akin to a forest microbiome on the immune systems of urban children (Roslund et al., 2020). This involved installing a forest floor condition in an outdoor play area of an urban day-care centre. Skin, stool and blood samples were taken from the children over the experimental period as they played in the space in order to monitor changes in their microbiome and immune markers. The study showed a measurable relationship between immune stimulation and micro-diversity of the space, which supports the general premise of the biodiversity hypothesis. Translocation of such niches into indoor environments might be feasible too, but would require research into the dynamics of such niches and their microbiomes when in an indoor environment.

Beyond landscape approaches and plant rewilding, little work currently exists on the role of bringing animals and farming back into cities specifically for the agenda of urban microbiome rewilding. In line with the farm effect described in Chapter 4, microbiota associated with farms have been positioned as beneficial for human immunoregulatory function. These strategies could be explored and expanded in line with research and real projects related to urban farms and vertical farming proposals in cities. These could also align with the agenda of food security amid the fragility of international food chains. Others have speculated on vertical dairy farming strategies for urban environments. Such approaches could contribute significantly to broader urban microbiome rewilding, though they may be limited in line with existing ethical and environmental concerns associated with the current dairy farming industry. Translocation of farm microbes to cities via microbial inoculants in a way that resembles microbiota from farm soils and dust may circumvent these issues.

## SYNTHETIC MICROBIAL COMMUNITIES

A further consideration we can expand on from the fields of landscape and rewilding approaches comes from ideas involving genetic engineering techniques to modify species, save endangered species or 'return' extinct species (Sherkow & Greely, 2013). Across broader challenges of climate change and food and energy shortages, synthetic biology is already positioned as one of the main transformative technologies needed to address these issues (Collins, 2012; Redford et al., 2013). Through the use of these tools it may be possible to engineer healthy microbes and communities, or even create new strains and communities with specific functions that could then be integrated into building materials. Bacterial species synthetically engineered to produce health-associated compounds have shown to be feasible in animal models in relation to diseases such as diabetes and obesity (Dou & Bennett, 2018; Duan et al., 2008). In the field of agriculture phytomicrobiome engineering explores the development of non-model bacteria and microbiomes to promote beneficial plant–microbe interactions (Ke et al., 2021). Architects in practice currently have little day-to-day engagement with synthetic biologists, though of course there is a precedent in the experimental field of biodesign. Designers in the field of living architecture are engaging with synthetic biology techniques to modify organisms in relation to material production(Dade-Robertson et al., 2015), optimising metabolic processes or editing new gene functions to produce new metabolic pathways to material production (Zolotovsky, 2017).

Through the contemporary disease paradigm of missing microbes and declining urban microbiodiversity, the science of genetic engineering could also be considered through the agenda of conservation. At the speculative end of this agenda, advocates of de-extinction are exploring the potential to resurrect previously lost species through genomic approaches using frozen DNA samples (Seddon et al., 2014). Notable examples include the proposed resurrection of the woolly mammoth to serve as an ecosystem engineer in Arctic tundra landscapes. The return of extinct species, especially larger mammals, in these cases often invokes polarised imaginaries that are either utopian or dystopian – the return of an old friend or an uncontrollable invasive entity from the past. Engineering extinct microbes or creating new microbes might raise similar imaginaries and dangers associated with fostering situations from which there is no return. The genetic modification of microbial strains for urban microbiome rewilding and human exposure is certainly feasible, but raises a mix of ethical and philosophical issues for the architectural field to consider.

## BIORECEPTIVE DESIGN

Looking beyond soft landscapes, the use of bioreceptive materials and designs for buildings could play a significant role in creating sources of microbiodiversity to rewild the urban environment. These materials can provide viable niches for the growth of microbial communities on their surface. The initial focus of the bioreceptive design agenda undertaken at the BiotA Lab was to utilise the vertical building envelope as a novel urban greening strategy. Driven mostly in line with a climatic discourse, bioreceptive materials were developed to support the growth of bryophyte species targeted not only for their photosynthetic agency but also for their ability to survive in urban environments even with limited moisture availability. These approaches in particular offer the potential for biodiversity growth on vertical surfaces on buildings and infrastructure that would otherwise be non-productive in terms of their biological agency. Less explored is the role that bioreceptive design approaches can play in rewilding the indoor microbiome. When applied on or near buildings, they offer significant potential to provide direct sources of environmental microbes for the indoor microbiome and human occupants.

Hypothetically, the same strategies can potentially contribute to creating more diverse building microbiomes. First, this would require research into how these approaches can be designed so as to facilitate indoor transport. This could be further augmented if the species and strains growing on these materials could be targeted to support holobiont-specific microbial diversity in line with the microbiome rewilding hypothesis. On one hand, bryophytes may already be well suited to this agenda. For example, mosses are known to harbour diverse microbial communities as part of their ecology – including bacteria, algae and fungi – which could potentially contribute to the broader beneficial agencies of microbiodiversity discussed in the previous chapters. On the other hand, by utilising a direct probiotic design approach, bioreceptive materials could also be inoculated with microbial communities or other living materials translocated from biodiverse sources (soil, forests and farms) which could potentially serve as direct sources of microbial diversity for buildings.

**MACRO-SCALE CHALLENGES**
The specific technical challenges for these approaches when applied to buildings lie in the difficulty of retaining sufficient moisture on vertical surfaces and harsh environmental exposure from the wind. The role of material porosity, along with three-dimensional surface geometries and forms, has been shown to slow water run-off and maintain moisture as well as providing microclimates of protection from wind that helped establish cryptogam growth. However, the complexities of these approaches in terms of geometry and the need for multiple material layers and porosity mixes brought their own challenges to their use in projects at the fabrication stage.

As these approaches have predominantly been explored using concrete casting into moulds, the majority of this research has been limited to 2½ dimensional panel-type geometries that are easily demouldable, and so has been limited to flat, non-structural façade panels (Cruz, 2022). Such practices have also been limited to single material application as craft-based, multi-material application approaches have proven inhibitory for industrial processes. Application on larger projects has also proven challenging, and the high costs of formwork and mould production has limited these to wall designs that rely on repeated patterning approaches. When applied at the scale of a building, these approaches lack sufficient variation to deal with building articulations or to provide sufficient variation to ensure sufficient microclimatic variations across and around a building.

In addressing these limitations and challenges, recent work has sought to explore these approaches through the lens of holobiont health. This involves bioreceptive design strategies, along with the probiotic methodologies developed in the previous chapters that align with the focus on informing indoor microbiodiversity. We sought to look conceptually beyond a façade panel approach as panels may be limited in their ability to impact the indoor microbiome as they are typically separated from the indoor environments by structural elements and moisture barriers. Instead, we sought to explore the design of building elements or components with the ability to create spatial forms for habitation as well the potential to inform the interior condition and start to shape human interactions with the biodiversity of the material matrix.

## THE PROBIOTIC PAVILION

We sought to explore this notion initially through a 1:1 prototype of an architectural design operating at the scale of a small pavilion. At this scale the architecture can engage with the environmental microbiome; it can also create a semi-indoor condition through which to explore and test the relationship between the outer and inner building fabric and the microbiome of the space. Research at this scale allows the physical prototype to serve as a spatial intervention to undertake a metagenomic study to learn about the microbiome of such an object in a real-world environment. The same studies could be done on much smaller material samples; however, the jump in scale was also important to begin to understand the realities and challenges of a probiotic design at the scale of buildings. This includes considerations of fabrication, transport, installation and costs.

The probiotic pavilion (Figure 5.1) is a small circular structure, large enough for 1–4 people to inhabit at a time (standing or sitting), aimed at offering a rich dose of microbial exposure to the inhabitants. Akin to a version of forest bathing, the microbes embedded in the materials provide the necessary microbial entanglements essential to the holobiont. Through mechanisms of touch, inhalation and ingestion, environmental soil microbes interact with the holobiont body's skin and airways while inhabiting the structure. It is imagined as a double-sided structure comprising multiple bioreceptive concrete mixes. The outer-facing surfaces will facilitate bryophyte species and their associated microbial communities, while the inside surfaces will be directly inoculated with soil microbes. These design approaches can be discussed in relation to the scales described in this book.

Figure 5.1 The probiotic pavilion

**Macro-Scale**

The geometry and form of the pavilion was simplified to a circular concept that defines a habitable space. The geometrical manipulation and orientation of this form seeks to exploit levels of exposure to surrounding biodiversity sources and other relevant environmental conditions to augment growth. Geometrical angles were determined according to conditions of sunlight/shade and prevailing winds from nearby green spaces. The side faces of the pavilion were tapered inwards at the top to provide inclines of less than 90 degrees to slow water run-off and maintain moisture. These then expand outwards before tapering inwards again at the bottom, creating stiffness via the double curvature.

**Meso-Scale**

Environmental conditions of shade and prevailing wind direction were then used to inform the meso-scale design aspects (Figure 5.2). The geometrical patterning was driven by environmental data on the orientation of the pavilion and by structural considerations. These patterns were created using a computational algorithm driven by these data sets; but they also create toolpaths for concrete extrusion during fabrication, thereby defining the physical niches in which the biological species will grow. These create protected environmental niches while maintaining structural integrity of the curved panels.

Macro-Scale: Probiotic Cities

These niches predominate mostly on the north and north-east faces that are in shade throughout the day. Where the surface angle is above 90° from the vertical, the toolpaths become tight and closed, offering structural stiffness, while on surface angles below 90° the toolpaths relax and open up, creating the growth zones for moss transplantation.

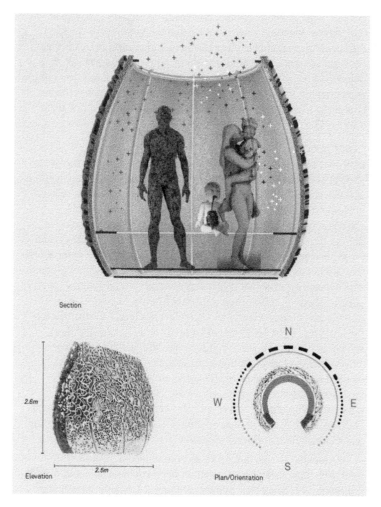

Figure 5.2 Probiotic pavilion section and meso-scale arrangement

## Micro-Scale

Using the same computational information, material mix and porosity were then determined by exposure. For outside faces, this was determined by areas on the pavilion surface simulated to receive wind flow via computational fluid dynamic (CFD) simulations based on geographical location. Material bioreceptivity on these outward-facing zones was optimised for bryophyte colonisation, with targeted porosity and specific concrete mix design. Although concrete is typically undesirable in terms of sustainability, the work aimed to pursue more sustainable alternatives to concretes that use ordinary Portland cement as the binder. Instead, novel concrete mixes were developed using blast furnace cement CEM III/B as it has greater sustainability, including resource conservation and energy saving compared to the Portland cement, and is also beneficial due to its lower pH.

## FABRICATION

In order to test the feasibility of the pavilion, the research sought to engage probiotic design with more contemporary industrial methods of fabrication in order to progress the work beyond the limitations of material casting approaches used previously. 3D concrete extrusion using industrial-scale robots allowed the production of double-curved, large-scale 3D building elements without the need for expensive moulds. It also facilitates the production of geometries that are not limited by the demoulding stage. Rather than traditional z-axis based approaches to printing derived through slicing, the new approach explored 3D printing the meso-scale defined toolpaths directly on to curved surfaces. Using this approach, formworks for the surfaces could be 3D printed quickly and cheaply and then reused to produce multiple components, with multiple variations. Afterwards the formwork material can be crushed and reused as hardcore. This permits the design of unlimited geometrical variations, allowing for application over larger areas without geometrical repetition. This approach may be better able to engage with the complexities of environmental deviations due to orientation, seasonal shading and wind direction.

This technique also facilitated an automated, multilayered approach that permits differentiation between the inside and outside condition. This process is shown in Figure 5.3, where the multiple layers comprised 3D-printed structural layers which were then infilled. Inside faces were infilled with fine porous probiotic materials

Figure 5.3 Pavilion fabrication sequence (clockwise from top left): formwork surface layer; layer 1 (inside-facing geometry); application of probiotic material between layers 1 and 2; application of carbon fibre sheets between layers; application of bioreceptive material layers 3–6; second application of carbon fibre sheets and extrusion of layers 7–10

for later inoculation with soil microbes. Carbon fibre meshes were inlayed into the middle layers during printing to offer additional structural reinforcement. Finally, outer-facing layers were infilled with a bioreceptive material, targeted for moss growth, before the final layers were printed, creating the moss niches.

**MACRO-SCALE BIOLOGICAL APPLICATION**
The final prototype, shown in Figure 5.4, was fabricated with industrial partner Incremental 3D as a test section of the larger pavilion which could then undergo metagenomic sequencing to study how the various microbial communities on the prototype

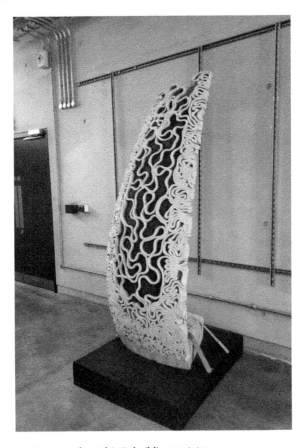

Figure 5.4 Macro-scale probiotic building prototype

behave and change over the course of a two-month experiment (which, at the time of writing, was still under way). Longitudinal sampling of prototype microbiomes in real-world environments will contribute to this understanding (Figure 5.5). Alongside learning about the architectural methodologies used, it is hoped we will learn about what happens specifically to the microbial communities on the surfaces. A big question for this research is how specific species will behave over time – whether they can remain or if other communities or species will eventually take over.

Biological application at large scales raises its own challenges. For the biological application of mosses, for example, we used a transplantation approach whereby mosses collected from a home-grown source were manually grafted into the niches on the front face of the pavilion. As well as the immediate aesthetic impact, for experimentation purposes this approach was used so that the

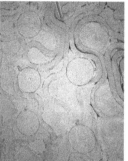

Figure 5.5 Macro-scale microbiological testing

microbial communities associated with the mosses were integrated into the prototype in time for the start of the sequencing study. While this was viable for the sake of the experimentation and may work in the case of small projects, manual transplantation of mosses on the scale of whole buildings would be unfeasible both because of the amount of moss required for transplant and the laborious and time-intensive process of transplantation. Another more viable method for bryophyte application is via inoculants to bioreceptive materials. Using this strategy means, however, that the visible appearance of growth will be slow and success is unpredictable, potentially taking years before colonies would be visibly established. These biological time frames are challenging to conventional ideas that a building is complete when construction ends and the keys are handed over. Instead these time-based and somewhat ephemeral conditions require a reframing of architectural conceptions of building aesthetics and permanence.

For the application of soil microbes to the inside faces, we used a technique of brushing on a liquid inoculant previously cultivated in the lab. Similarly with the application of microbial inoculants, more biologically intelligent approaches may be required for large-scale applications. As well as the laborious process of application, the cultivation of large amounts of liquid inoculants is costly and challenging. In upcoming work we will be exploring the potential to directly integrate bacterial and moss spores into the material mixes during the fabrication stage. In this way, the biological agency could be integrated into the process rather than applied later. Research will be needed to determine whether these spores can survive the physical and chemical stresses associated with this methodology, and thus to understand their agency.

**PROBIOTIC BUILDINGS**
In the final part of this chapter I want to speculate on broader architectural methodologies for how we might design buildings that can engage fully with a probiotic framework. This looks beyond specific material applications and façade designs towards a fully bottom-up approach to planning whole buildings, their programme and their form so that the strategies to maximise biodiversity discussed in this book are fundamentally embedded within each stage of the building design. In this way, maximising

microbiodiversity is positioned as a fundamental driver of the building design rather than something added or retrofitted later.

At the macro-scale, buildings cannot be designed in isolation from their site and surrounding urban environmental conditions. Designs must respond then to the site-specific ecological and environmental parameters and, importantly, in a way that informs the microbiome in line with existing knowledge. As we have learned, these conditions are relational and time based, and so require methodologies that are adaptable. In line with ecologists – who develop 'ecological network models' for rewilding projects to predict how ecologies emerge from the configuration of new species interactions – microbiologists, building engineers and architects would need to develop 'building–microbe network models' that can be adapted within a more symbiotic framework to develop beneficial networks. These approaches will require methodologies to incorporate and interpret the large amounts of microbial information that come from metagenomic sequencing and environmental data. Architects are well placed to inform these models with existing knowledge of spatial simulation and interaction modelling techniques, for example. The development of these models will then allow for their use by architects without specific knowledge of this scientific field.

Within my research cluster (RC7), we have been exploring the development of such models using computational approaches. The methodologies developed are centred on the understanding that individual, one-off buildings will be unable to contribute significantly to creating truly biodiverse built environments. Building–microbe network models instead can inform multiple building designs in urban areas to facilitate broader strategies that can contribute to landscape connectivity and microbiodiversity corridors. The projects also explore how these models can help engage the diversity of stakeholders needed to plan and implement these approaches. Hence, the design focus lies in the development of probiotic design 'platforms' that can be used not only by architects but also by local authorities, planners and other built-environment stakeholders. In this way the platform becomes a tool to facilitate multidisciplinary engagement that embeds continuous agency for non-human life throughout the design process, irrespective of individual agendas. Importantly, these platforms also allow actors from non-design based disciplines (such as microbiologists or microbial ecologists) to engage with the design through implementation of specific data sets which can be embedded into the platform to inform the building.

## MICROBIOME DATA SETS

Fundamental to this is exploration of how probiotic buildings could be designed and planned in relation to these data sets, along with other information relating to the surrounding green spaces and wind directions associated with biodiverse sources for a given site. In the BioHealer.AI project, we sought to develop a computational process for integrating some of this microbially important environmental information into a platform that would inform the design of a building for a specific site in a way that maximises exposure and capture of airborne microbial diversity from sources of biodiversity within a certain radius.[1] This involved the use of multiple data sets to train machine learning (ML) models which could then be used to inform the building.

In the first step, an image-recognition based ML model was trained using generative adversarial networks (GANs) to detect and measure green spaces from aerial images of any given city (Figure 5.6). This was positioned as a kind of 'urban diagnosis' stage to understand a site in terms of how much green space is within a defined locale. 'Unhealthy' sites would be characterised by low amounts of surrounding green space, while 'healthy' sites would contain large amounts. Using this ML model, any building plot in a given geographical location and site could then be 'zoned' according to its relative distance from those green spaces. Sites located furthest from green spaces would require more 'direct' probiotic design strategies to inform a diverse indoor microbiome, while building plots closer to natural spaces might inform buildings that could rely on more indirect strategies. Such an approach could potentially inform existing local planning policies and design guidelines whereby building projects could be assigned a level of 'microbiodiversity' they need to achieve, similar to how energy targets or sustainable drainage targets are specified. This could also relate to the notion that humans have a 'right' to beneficial microbial exposure in a similar way that a 'right to light' exists as an embedded condition of building design.

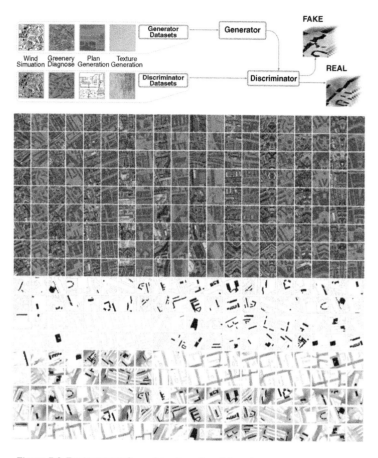

Figure 5.6 Environmental, machine learning data sets

Building on this zoning approach, a second model was then trained according to wind data which could then be calculated for the site and its surroundings. Air samples taken downwind from natural areas have demonstrated measurably greater diversity of microbial communities compared to upwind samples, dominated by those from plant and vegetative sources (Lymperopoulou et al., 2016). Using particle simulations and point cloud data within procedural design software, this information can be imported to a digital model and used directly to inform an initial building mass

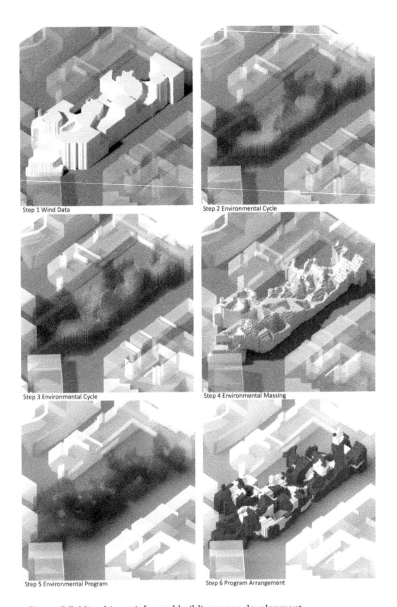

Step 1 Wind Data

Step 2 Environmental Cycle

Step 3 Environmental Cycle

Step 4 Environmental Massing

Step 5 Environmental Program

Step 6 Program Arrangement

Figure 5.7 Microbiome-informed building mass development

Macro-Scale: Probiotic Cities

that is geometrically sculpted and optimised (alongside other more traditional architectural concerns, such as context) for maximum exposure to and 'capture' of airborne microbes from the green spaces.

This data can be calculated at different vertical heights to inform the development of an initial mass, here applied to a mid-rise, mixed-use building typology shown in Step 1 of Figure 5.7. After this initial mass is created, other environmental data sets can then be incorporated, and could begin to further sculpt and shape the overall building mass to inform various niche conditions for green biodiversity which could be planned into the building form towards an architecture where mass and form are driven by the notion of exposure. In Steps 2–4 of Figure 5.7, solar, thermal and moisture information from multiple environmental cycles were simulated to inform a variety of biodiversity niches, including managed and unmanaged conditions.

Vertical surfaces in shaded areas not receiving high solar gain were parametrically sloped and assigned bioreceptive materials to create large surface areas of unmanaged cryptogam growth, including algae, lichens and mosses. Vertical surfaces receiving more sunlight were assigned green vertical planting with larger plant species which can be then spatially aligned with the fenestration to ensure that the biodiversity associated with green walls is close to windows so that microbes may be transported inside. Horizontal surfaces were also considered in terms of their environmental exposure and assigned different functions of green spaces. Rejecting aesthetic drivers of green spaces as nice places to have, these spaces were designed to promote interactions between human and environmental microbiomes. They could include multiple typologies of green space, such as allotments for vegetable growing, garden spaces for diverse plant species or more wild areas lining circulation routes. The same data sets were then used to begin to inform the floor plan of the building. Live work spaces were spatially organised to ensure locality to areas of high biodiversity to promote microbial entanglements.

### ARCHITECTURAL AGEING AS A FACTOR IN MICROBIODIVERSITY

While there may be no specific aesthetic that defines what probiotic buildings might be or look like, the concept of maximising

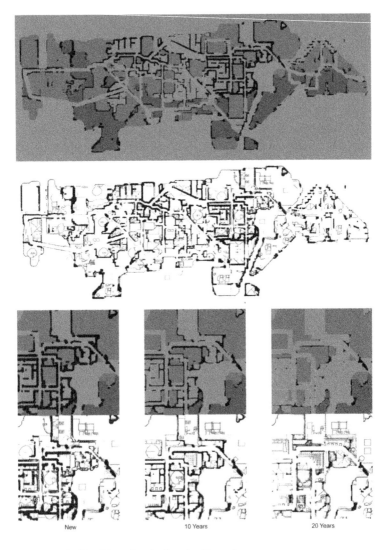

New          10 Years          20 Years

Figure 5.8 Microbiome-inspired ageing floor plans

microbiological diversity and growth on and inside a building will result in conditions that challenge the clean aesthetics of modernism and the sterile aesthetics of the contemporary globalised city. While buildings with integrated vegetation have been preferred and considered more beautiful than buildings without (White &

Figure 5.9 BioHealer.AI building proposal

Gatersleben, 2011), these have mostly related to green and plant wall vegetation. Unplanned biocolonisation has more traditionally been associated with building failure, though positive aesthetics of biological colonisation on buildings has been discussed through notions of romanticism and nostalgic aesthetics of aged castles or follies (Cruz & Beckett, 2016). Important to these conditions is the role of ageing, in terms of both the physical and the chemical properties of materials. As materials age, they tend to become more bioreceptive. Weathering and erosion over long time periods facilitate conditions of surface roughness and porosity, while chemical conditions such as pH levels slowly reduce. This ageing also reshapes the architecture – hard corners become rounded, pits and crevices are shaped by wind and water. In the extreme, materials disappear.

The project explored this concept of designing a building to purposely age, both in its material and physical condition, as a speculative design strategy to maximise microbiodiversity in/on buildings. Accelerated ageing of materials was explored through the carbonation of concrete materials to lower the pH (Chang & Chen, 2006), while the use of low compacted porous materials on non-structurally relevant parts of buildings sought to facilitate physical erosion as a precursor for colonisation and to instigate aged aesthetics. Even more speculatively was that this notion of erosion could be extended to the programme and spaces in the building. In Figure 5.8 the platform informs areas of material erosion and decay. As parts of the building were allowed to erode, inside spaces slowly became outside spaces, where the microbiodiversity of the building is able to flood deeper into the floor plan; and the resultant building aesthetic, shown in Figure 5.9, is one where the aesthetic of ageing is embraced as a fundamental part of the building tectonic.

**NOTE**

1  BioHealer.AI was a project undertaken in RC7 in 2021/22 with Yuqian Gao, Xiaoying Fu, Hangchuan Wei and Yuhan Wu.

**REFERENCES**

Chang, C.-F., & Chen, J.-W. (2006). The experimental investigation of concrete carbonation depth. *Cement and Concrete Research, 36*(9), 1760–1767.

Collins, J. (2012). Synthetic biology: bits and pieces come to life. *Nature, 483*(7387), S8–S10.

Cruz, M. (2022). *Poikilohydric living walls.* Bartlett Design Research Folios. Bartlett School of Architecture. bartlettdesignresearchfolios. com.

Cruz, M., & Beckett, R. (2016). Bioreceptive design: a novel approach to biodigital materiality. *Architectural Research Quarterly, 20*(1), 51–64. https://doi.org/10.1017/S1359135516000130.

Dade-Robertson, M., Figueroa, C. R., & Zhang, M. (2015). Material ecologies for synthetic biology: biomineralization and the state space of design. *Computer-Aided Design, 60*, 28–39. https://doi. org/10.1016/j.cad.2014.02.012.

Dou, J., & Bennett, M. R. (2018). Synthetic biology and the gut microbiome. *Biotechnology Journal, 13*(5), 1700159.

Duan, F., Curtis, K. L., & March, J. C. (2008). Secretion of insulinotropic proteins by commensal bacteria: rewiring the gut to treat diabetes. *Applied and Environmental Microbiology, 74*(23), 7437–7438.

Hui, N., Grönroos, M., Roslund, M. I., Parajuli, A., Vari, H. K., Soininen, L., Laitinen, O. H., Sinkkonen, A., & ADELE Research Group. (2019). Diverse environmental microbiota as a tool to augment biodiversity in urban landscaping materials. *Frontiers in Microbiology, 10*, 536.

Ke, J., Wang, B., & Yoshikuni, Y. (2021). Microbiome engineering: synthetic biology of plant-associated microbiomes in sustainable agriculture. *Trends in Biotechnology*, *39*(3), 244–261.

Lymperopoulou, D. S., Adams, R. I., & Lindow, S. E. (2016). Contribution of vegetation to the microbial composition of nearby outdoor air. *Applied and Environmental Microbiology*, *82*(13), 3822–3833.

Meadow, J. F., Altrichter, A. E., Kembel, S. W., Kline, J., Mhuireach, G., Moriyama, M., Northcutt, D., O'Connor, T. K., Womack, A. M., Brown, G. Z., Green, J. L., & Bohannan, B. J. M. (2014). Indoor airborne bacterial communities are influenced by ventilation, occupancy, and outdoor air source. *Indoor Air*, *24*(1), 41–48. https://doi.org/10.1111/ina.12047.

Mills, J. G., Bissett, A., Gellie, N. J. C., Lowe, A. J., Selway, C. A., Thomas, T., Weinstein, P., Weyrich, L. S., & Breed, M. F. (2020). Revegetation of urban green space rewilds soil microbiotas with implications for human health and urban design. *Restoration Ecology*, *28*, S322–S334.

Redford, K. H., Adams, W., & Mace, G. M. (2013). Synthetic biology and conservation of nature : wicked problems and wicked solutions. *PLOS Biology*, *11*(4), e1001530. https://doi.org/10.1371/journal.pbio.1001530.

Robinson, J., Watkins, H., Man, I., Liddicoat, C., Cameron, R., Parker, B., Cruz, M., & Meagher, L. (2021). Microbiome-inspired green infrastructure (MIGI): a bioscience roadmap for urban ecosystem health. *Preprints.org*. 2021040560. https://doi.org/10.20944/preprints202104.0560.v1.

Roslund, M. I., Puhakka, R., Grönroos, M., Nurminen, N., Oikarinen, S., Gazali, A. M., Cinek, O., Kramná, L., Siter, N., & Vari, H. K. (2020). Biodiversity intervention enhances immune regulation and health-associated commensal microbiota among daycare children. *Science Advances*, *6*(42), eaba2578.

Seddon, P. J. (2010). From reintroduction to assisted colonization: moving along the conservation translocation spectrum. *Restoration Ecology*, *18*(6), 796–802.

Seddon, P. J., Moehrenschlager, A., & Ewen, J. (2014). Reintroducing resurrected species: selecting deextinction candidates. *Trends in Ecology & Evolution*, *29*(3), 140–147.

Sherkow, J. S., & Greely, H. T. (2013). What if extinction is not forever? *Science*, *340*(6128), 32–33.

White, E. v., & Gatersleben, B. (2011). Greenery on residential buildings: does it affect preferences and perceptions of beauty? *Journal of Environmental Psychology*, *31*(1), 89–98. https://doi.org/10.1016/j.jenvp.2010.11.002.

Zolotovsky, K. (2017). *Guided growth: design and computation of biologically active materials*. MIT Press.

# Conclusion

## Chapter 6

DOI: 10.4324/9781003207917-6

The emerging sciences of the microbiome are radically impacting many fields, yet specific medical knowledge of the relationship between the microbiome and health remains incomplete. The same is true of interactions between humans and the microbiomes of the built environment; and so, currently, practical applications for creating healthy building microbiomes remain a future vision (National Academies, 2017). While this research predominates in the medical and building science fields, this book hopes to make visible the important role that architecture can play in shaping this vision. As discussed, this will require a shift from antibiotic mentalities to probiotic mentalities. In Chapter 2, I set up the need for this recalibration of the way we manage life in buildings based upon contemporary human–microbe relationships. We read how these can be underpinned by microbiome science and the figure of the human holobiont that makes visible the beneficial role that informed indoor microbiomes can play in health.

Architects can contribute much beyond the current focus on the characterisation and identification of microbiomes in existing buildings. They can also look beyond the narrow focus on indirect strategies identified in the current literature. Architects and biodesigners, through their very nature of designing with life, can inform radically new building designs using biologically active materials and components that will shape radically different indoor microbiomes. These can have profound benefits for health if they are targeted and designed in a manner that will facilitate beneficial entanglements with the human microbiome. In Chapter 4, I used the term 'entanglements' in order to reframe microbial transmissions from a problematic condition to one that is beneficial. Architecture has a key role in shaping these entanglements, not only through the constitution and spatial arrangement of microbes in a building but also through informing how these entanglements are established and accessed across cities.

Equal access and exposure to this beneficial microbial diversity will be important if probiotic cities are to have equitable impacts on public urban health. In the current city, access to green space is not always equal; and, when considered through the lens of holobiont health and spatial justice, it is clear that access to microbes is not equal either. Disadvantaged and minority groups are less likely to have access to high-quality, biodiverse, green and blue spaces in urban areas as a result of their higher representation in low socio-economic areas (Roe et al., 2016; Stephens, 2015). This social inequality in access to macrobiodiversity will subsequently equate to reduced access to microbiodiversity and its discussed benefits for holobiont health. In some cases, these microbial health disparities are exacerbated by the fact the people of lower socio-economic status are more likely to be exposed to harmful microbes in the built environment. These include conditions in buildings with water damage and building deterioration, which can contribute to harmful microbial exposure associated with sick building syndrome (Abdul-Wahab, 2011).

While the recent Covid pandemic reminded us of the need to reduce exposure to pathogens in the built environment, there is a risk that it may also exacerbate antibiotic mentalities in the built environment – which will only contribute further to the contemporary pathologies of antimicrobial resistance (AMR) and chronic immune diseases. As we seek to subvert this trend, probiotic materials and buildings have the potential to facilitate targeted beneficial entanglements to improve human health.

## REFERENCES

Abdul-Wahab, S. A. (ed.) (2011). *Sick building syndrome in public buildings and workplaces*. Springer.

National Academies of Sciences, Engineering, and Medicine (2017). *Microbiomes of the built environment: a research agenda for indoor microbiology, human health, and buildings*. National Academies Press.

Roe, J., Aspinall, P. A., & Ward Thompson, C. (2016). Understanding relationships between health, ethnicity, place and the role of urban green space in deprived urban communities. *International Journal of Environmental Research and Public Health, 13*(7), 681.

Stephens, C. (2015). The indigenous experience of urbanization. In P. Grant (ed.), *State of the world's minorities and indigenous peoples, 2015* (pp. 54–61). Minority Rights Watch International.

# Index

Note: *Italic* page numbers indicate figures.

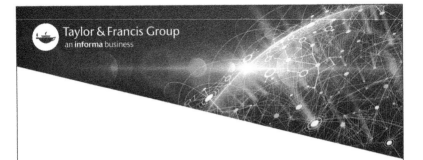